INTERPRETATION OF NMR SPECTRA

An Empirical Approach

INTERPRETATION OF NMR SPECTRA

An Empirical Approach

by

Roy H. Bible, Jr., Ph. D.

G. D. Searle & Co.

PLENUM PRESS
NEW YORK
1965

First Printing – February 1965
Second Printing – September 1965
Third Printing – March 1966 (with corrections)

Library of Congress Catalog Card Number 64-20741

Preface

In writing this book I had two main objectives: (1) to teach the organic chemist how to interpret proton magnetic resonance spectra, and (2) to provide the reference data which are constantly needed in the use of proton spectra. I have felt that it was important to point out not only the information which can be gained from spectra, but also the limitations and the potential pitfalls. All of the important facts are organized into tabular summaries. Every effort has been made to present the material clearly, concisely, completely, and accurately. At the same time, subjects not directly related to the interpretation of spectra have been omitted. Thus, while the conclusions drawn from theory are presented, the theory itself has been avoided.

There are a number of advantages in learning the empirical facts before learning the theory. First of all, in interpreting spectra one usually has to rely on his knowledge of the accumulated empirical correlations much more than on his knowledge of the theory. In fact, one could know all of the theory and still not be able to interpret spectra unless he also knew the empirical facts. Secondly, the theory is much more easily understood after the facts have been mastered.

This book began as a seminar for my colleagues at G. D. Searle & Co. The objective of that seminar was to present all of the fundamentals needed in the interpretation of proton spectra. That material, which constitutes Chapter 2 of this book, was later assembled as a booklet for use in conjunction with a workshop presented at Searle by Dr. Harmon W. Brown, Jr., and Dr. Donald P. Hollis of Varian Associates.

My functions in writing this book have been primarily to collect, evaluate, organize, and present the material in terms familiar to the organic chemist. I have drawn heavily from the texts by L. M. Jackman; J. A. Pople, W. G. Schneider, and H. J. Bernstein; K. B. Wilberg, and B. J. Nist; and J. D. Roberts. In addition, I gained much information from the series of lectures at Searle by Dr.

Harmon W. Brown, Jr., and Dr. Donald P. Hollis of Varian Associ-
ates, and from lectures elsewhere by Dr. Frank A. L. Anet, Dr.
Kenneth W. Bartz, Dr. Wallace S. Brey, Jr., LeRoy F. Johnson,
Dr. Paul C. Lauterbur, Dr. John A. Pople, Dr. John D. Roberts,
Dr. James N. Shoolery, and Dr. George Slomp.

A number of points were clarified in discussions with Dr. Fred
Kaplan, Dr. Thomas J. Flautt, and Dr. Frank A. L. Anet.

My colleagues at Searle have contributed to this book in many
ways. They have brought much literature to my attention, pointed
out many unusual features in spectra of their own compounds, and
have challenged me with many questions.

The unique research environment at Searle has been important
both in my own study of NMR and in the development of this book.
I am particularly indebted to my supervisors, Dr. Robert R. Burtner,
Dr. Byron Riegel, and Dr. Albert L. Raymond, for their support in
this project.

Most of the spectra used were determined by Miss Diana Ede
under the direction of Aristides J. Damascus, supervisor of the
spectral laboratory at Searle. The spectral laboratory is part of
the Analytical Department, which is administered by Dr. Robert T.
Dillon. The enthusiasm, ability, and support of this group has been
essential to this book.

The graphical treatment of the ABX system given in Chapter 4
was worked out in collaboration with David W. Calhoun, who served
as my principal consultant in mathematics.

I am especially indebted to my wife, Harriett, and to Edward
A. Brown and Dr. Thomas J. Flautt for their careful reading of the
manuscript. Their criticisms have led to considerable improve-
ments in all aspects of this text.

In addition to being my chief critic, my wife also shouldered
many responsibilities which permitted my completion of this book
in a reasonable length of time.

I am grateful to Dr. Kenneth B. Wiberg, Dr. Bernard J. Nist,
and W. A. Benjamin, Inc., for permission to reproduce the material
given in Figure 4-4, and to Varian Associates for permission to
reproduce a number of spectra.

The original draft of Chapter 2 was typed by Mrs. Karen
Khubchandani. The present manuscript was typed by Mrs. Diane
George and Mrs. Sandra Blume. Many of the structures were drawn
by Mrs. Blume, who did much of the art work in the original draft
of Chapter 2. I have also been aided during the development of this

book by the photographic work of Clifford Kornoelje and the multilith work of George Wolfram and Phillip Sosnowski.

The efficient and pleasant staff of Plenum Press did much to improve the presentation of the material.

Criticisms of this book will be appreciated. In particular, comments concerning residual errors or the emphasis placed on specific topics will be helpful in the preparation of possible future editions.

Roy H. Bible, Jr.

Contents

Scope, Limitations, and Applications of Nuclear Magnetic Resonance

Approximately half of the known isotopes have nuclei which behave like spinning magnets. When placed in a magnetic field, these nuclei can be caused to absorb radio-frequency energy. In the most favorable cases, the small amount of energy which is absorbed can be measured as a function of the strength of the magnetic field. In Table 1-1 the nuclei of interest to organic chemists are ranked according to their usefulness in nuclear magnetic resonance (NMR) studies.

There are three factors which determine the ranking of nuclei in Table 1-1:

(1)   the strength of the nuclear magnet,
(2)   the relative natural abundance of the isotope, and
(3)   the evenness with which the nuclear charge is distributed over the surface of the nucleus.

The most useful nuclei behave like strong magnets, are relatively abundant, and have a spherical distribution of nuclear charge.

For most of these nuclei the general technique is to place a nonviscous, concentrated solution of the compound (or the pure compound if it is a liquid) in a strong, homogeneous magnetic field. The sample is then irradiated with a radio-frequency signal. The intensity of absorption of the radio-frequency signal is plotted as a function of increasing magnetic field strength.

It can be seen from Table 1-1 that of the three most common isotopes in organic compounds, C^{12}, H^1, and O^{16}, only H^1 nuclei (protons) can give usable NMR signals. Actually, the nuclei of C^{12} and O^{16} are nonmagnetic. This is, in a way, quite fortunate, for it results in much simpler proton spectra. A number of the other common isotopes, notably F^{19}, P^{31}, and N^{14}, are magnetic. While these other nuclei do affect the proton spectra, they usually occur in relatively small numbers in any one compound.

By far, most of the studies in NMR have been concerned with

TABLE 1-1

Usefulness of Common Nuclei in Natural Abundances
in NMR Studies

Most useful	H^1	F^{19}	P^{31}		
Useful	B^{11}	N^{14}	Si^{29}		
Usable	H^2	C^{13}	N^{15}	O^{17}	S^{33}
Not usable	C^{12}	O^{16}	Si^{28}	Si^{30}	S^{32}
	Cl^{35}	Cl^{37}	Br^{79}	Br^{81}	I^{127}

protons. This has resulted from the interest in organic compounds, the high sensitivity of protons to this technique, and their high natural abundance. These factors have also made it commercially feasible to produce, at moderate costs, easily operated spectrometers for proton studies. With present commercial instruments, good proton magnetic resonance spectra can be obtained with almost the same ease and rapidity as IR spectra. An instrument for routine examination of F^{19} spectra has recently been introduced. Investigations of P^{31} are somewhat more difficult, while investigations of other nuclei are often quite difficult.

The magnetic field strength at which a nucleus absorbs energy of a particular radio frequency is primarily determined by the nucleus. Figure 1-1 shows the approximate magnetic field strength at which various nuclear isotopes absorb energy of 60 Mcps frequency. To a very small extent, the magnetic field strength at which a nucleus absorbs energy of a particular radio frequency is dependent on the chemical (or more exactly, the electronic) environment of the nucleus. Thus protons in different locations in a molecule absorb a 60-Mcps signal at slightly different magnetic field strengths. Were it not for this slight dependence on molecular environment, the NMR technique would be of little importance. Spectra which reveal these small differences in chemical environments are called "high-resolution spectra." Present-day instruments show even finer details and are, in a sense, "ultrahigh-resolution" spectrometers.

A concept of the relative magnitude of differences due to different nuclear species and the much smaller differences due to different molecular environments can be gained from the distance scale in Figure 1-1. If the total range of the signals due to protons in all environments were spread over a chart 20 in. long, the chart would have to be extended for over 2 miles in order to detect the nearest other nucleus (F^{19}). On this scale, many intimate structural details are clearly revealed by differences in chart positions of less than $\frac{1}{10}$ in.

Under proper experimental conditions, the intensity of absorption of the radio-frequency signal is proportional to the number of nuclei causing the absorption. Thus, the spectrum not only reveals the number of different environments in which protons are located, but also the relative number of protons of each type.

The presence of other nearby magnetic nuclei in the molecule can give rise to finer details in the spectrum. These finer details are, to a great extent, predictable, thus revealing even more structural information.

In addition, the rates at which nuclei are undergoing change in their environment can often be determined from the spectra.

The types of information obtainable from high-resolution spectra are summarized in Table 1-2.

For high-resolution spectra, the sample must be studied in a nonviscous liquid state. Solids give only very broad bands and can only be examined on a different ("wide line") spectrometer. The

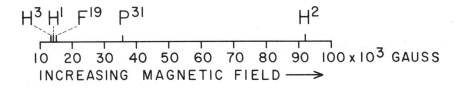

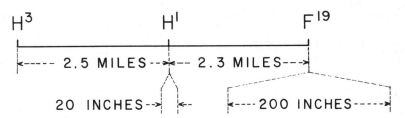

Fig. 1-1. Magnetic field strength at which common nuclei would absorb energy from a 60-Mcps signal. The lower chart indicates the relatively small region over which individual nuclear signals occur compared to the differences among different nuclei.

broad bands can be used to determine some useful information, but the amount of information is quite limited [11].

The nuclei being studied must be present in large numbers in the solution. The usual concentration required for proton spectra is 5 to 20%. Since the number of suitable solvents is limited, this concentration requirement is a limitation of the use of the technique. The amount of compound required for proton spectra ranges from 2.5 mg to 30 mg. In some cases, this sample requirement also presents problems. Special ways of circumventing these difficulties are presented in the next chapter. The sample solubility requirements are summarized in Table 1-3.

There are four general types of problems which may be solved by the use of NMR spectra:

(1) structural determinations and verifications,
(2) estimation of purity of samples,
(3) analysis of mixtures, and
(3) the rates of some dynamic processes.

The greatest use of NMR has been for the determination and verification of molecular structures. This could be easily surmised from the types of information obtainable (Table 1-2).

A rather surprising secondary use of proton spectra arises from the sensitivity of the band positions to very slight changes in

TABLE 1-2

Types and Origin of Information Obtainable from High-Resolution Spectra

Information obtainable	How obtained
Number of groups of nuclei in different molecular (electronic) environments	From number of absorption bands
Relative number of nuclei in each different environment	From relative areas under the absorption curve for the various bands
Relationship of one group of nuclei to other nearby groups of nuclei	From finer details in the absorption bands
Some rates at which nuclear environments are being changed	From number and shapes of absorption bands; often studied by causing changes in the rate

TABLE 1-3

Approximate Limitations for Proton Spectra Using a Single
Scan on the Varian A-60 Instrument
(The HR-100 is four times as sensitive.)

Compound Requirements	
Amount	2.5 to 30 mg
State	Must be in a nonviscous liquid state
Solubility requirement	5 to 20%

the molecular environment of the protons. Quite often, mixtures of
closely related compounds are difficult to distinguish from single
pure compounds. The mixture may have a sharp melting point and
be unresolved on thin-layer chromatography. Often neither the IR
nor UV spectra display absorption bands which indicate the number
of components. The proton spectra of many such mixtures immediately show, by the number of different types of protons and the
relative numbers of these protons, that the substance cannot be a
single compound.

The exact manner in which information can be extracted from
proton spectra using an empirical approach will be considered next.

Chapter 2

Fundamentals of the Interpretation of Proton Spectra

Although the early workers in nuclear magnetic resonance placed the subject on a firm theoretical basis, they simultaneously placed a number of stumbling blocks, such as electric quadrupole moments and diamagnetic susceptibilities, in the path of the average organic chemist. Many organic chemists effectively utilize IR and UV spectra without being concerned, and in some cases without knowing, the true nature of the particular transition which is involved. It is the purpose of this chapter to show how NMR spectra can be used in the same fashion.

This discussion can be conveniently divided into two main sections: (1) general remarks and (2) interpretation of spectra. Under general remarks, a brief discussion of the technique required will be given. A few statements must be made concerning the theory, and the characteristics of a good spectrum will be noted. In order to interpret the spectra, three aspects will be examined: band positions, band intensities, and multiplicity of peaks in these bands.

This discussion will be oriented primarily to spectra obtained at 60 Mcps, such as those obtained on the Varian A-60, the instrument available to most organic chemists.

If the microcell is used with the Varian A-60, about 2.5 mg of material is sufficient for an NMR spectrum. If the more commonly available macrocell is used, about 30 mg is required. The compound must be dissolved to the extent of about 10% in a suitable solvent. This means that the 2.5-mg sample must be dissolved in 0.025 ml of solvent, and the 30-mg sample must be dissolved in 0.3 ml. By far the best solvent, theoretically, for NMR spectra is carbon tetrachloride. Unfortunately, however, many organic compounds are not sufficiently soluble in carbon tetrachloride. As a consequence, the most widely used solvent is deuterochloroform ($CDCl_3$). All but one (Figure 5-2) of the spectra in this book were determined using deuterochloroform as a solvent. If the compound is not soluble to the required 10% in deuterochloroform, other deuterated solvents

TABLE 2-1

Common NMR Solvents

The dark areas are not usable under the conditions normally employed. Some of the absorption bands are due to impurities which vary in concentration.

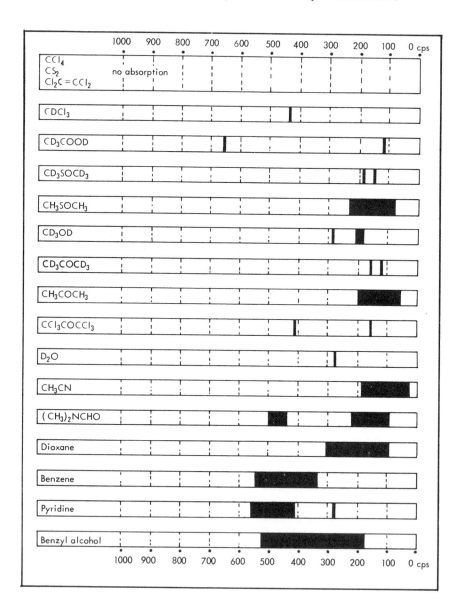

may be employed. For the examination of limited portions of the spectrum, some proton-containing solvents may be used. It must be remembered, however, that there are solvent effects. Some common solvents are given in Table 2-1 together with the spectral regions in which they can be used.

There are two methods by which a usable spectrum can be obtained using less than 2.5 mg of compound (see Table 2-2). The first is to employ a more sensitive spectrometer. For example, a fourfold increase in sensitivity can be obtained using the Varian HR-100. The second method is to repeatedly scan the spectrum and to analyze the results using a computer. In this way the background noise is averaged out and the sensitivity is increased by the square root of the number of scans. Thus, if the spectrum is repeated 16 times, a fourfold increase in sensitivity is obtained. This technique involves a computer of average transients, and the computer used is called a "C.A.T." Currently available C.A.T.'s can be made to automatically repeat the desired portion of the spectrum.

If the compound is not soluble to the required 10% in a suitable solvent, there are four methods which may give a usable spectrum (Table 2-2): A more sensitive spectrometer can be employed, the spectrum may be repeatedly scanned and the results analyzed with a C.A.T., or the spectrum can be determined in a different solvent or at an elevated temperature.

In the NMR spectrum the intensity of absorption of radio-frequency energy of approximately 60 Mcps frequency is plotted as a function of increasing magnetic field strength (see Figure 2-1).

TABLE 2-2

Special Methods of Obtaining Usable Spectra with a Small Amount of Compound or an Insoluble Compound

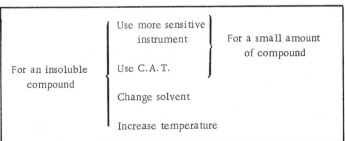

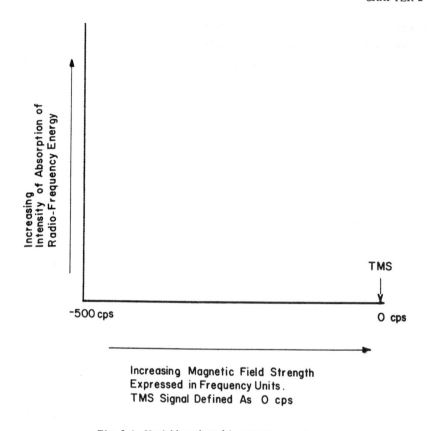

Fig. 2-1. Variables plotted in a proton spectrum.

The magnetic field strength is most easily measured in frequency units (cps). Changes of less than one part in 60 million in the strong (about 14,000 gauss) magnetic field are significant in the spectrum. As is easily imagined, measurement of this enormous field strength, in absolute terms, to the required accuracy would be quite difficult. So rather than measure the field strength on an absolute basis, a reference compound, tetramethylsilane [$(CH_3)_4Si$; TMS] [76], is placed in the solution.   All of the protons in tetramethylsilane absorb at the same magnetic field strength, producing a sharp, single peak. Thus all field strengths are measured in frequency units with reference to the field strength at which the tetramethylsilane signal (defined as 0 cps) is observed.   The difference in position of two sharp, single signals in the spectrum is called the difference in chemical shifts. Since it is understood that all signal positions are relative, the position of a signal with reference to a standard is often simply called the chemical shift.

A reference compound dissolved in the solution is referred to as an "internal" reference. Since tetramethylsilane is insoluble in D_2O, the reference compound must, in this case, be sealed in a separate capillary tube. The capillary tube is then placed directly in the test tube which holds the D_2O solution. A reference compound contained in a separate capillary tube is called an "external" reference. Sodium 2,2-dimethyl-2-silapentane-5-sulfonate (DSS), $(CH_3)_3SiCH_2CH_2CH_2SO_3Na$, is a useful water-soluble (internal) reference compound [69].

The compound to be examined is thus dissolved to the extent of about 10% in deuterochloroform which has been made up to contain approximately 1% tetramethylsilane. This solution is placed in a special test tube which is then placed in a strong magnetic field. The tube is spun about its long axis to help average out any differences in the magnetic field strength. The absorption of radio-frequency energy of about 60 Mcps frequency is measured as a function of increasing magnetic field.

The results of this determination are shown in Figure 2-2. There are three traces shown on this graph. The two lower traces are absorption curves, while the upper trace, which consists of a series of plateaus, is an integration curve. The integration curve will be discussed later. In the two lower curves, the intensity of absorption of radio-frequency energy is plotted as a function of increasing magnetic field strength expressed in frequency units. It will immediately be noted that, while the magnetic field is increasing from left to right, the numbers on the graph decrease from left to right. These are actually negative numbers, the magnetic field strengths all being less than the field strength at the reference signal of tetramethylsilane.

With a 60-Mcps signal and a 14,000-gauss magnetic field (such as employed on the Varian A-60), only protons (nuclei of hydrogen) and noise show up in the spectrum. No other nucleus, functional group, or even nucleus of a hydrogen isotope gives rise to absorption in the spectrum. In order to see absorption due to the next nucleus (F^{19}) which would show up at a higher field, the graph, as it is usually run, would have to be extended 2.3 miles to the right. The only nucleus at a lower field would be tritium (H^3), which would be 2.5 miles to the left (see Figure 1-1).

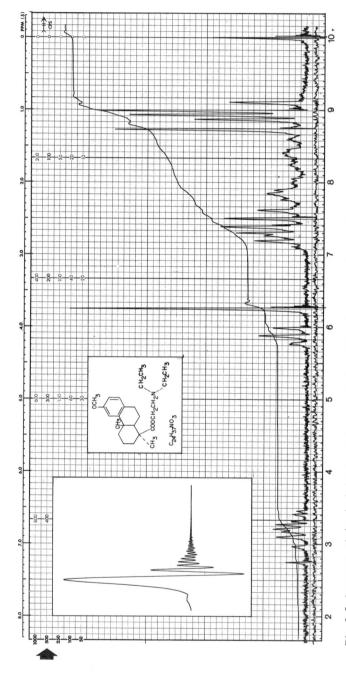

Fig. 2-2. Spectrum of the diethylaminoethyl ester of O-methylpodocarpic acid in CDCl$_3$. The bottom trace is the absorption curve from −1000 to −500 cps. The middle trace is the absorption curve from −500 to 0 cps. The trace which appears as a series of plateaus is the integration curve.

Most protons in organic compounds show absorption bands within the region of −600 cps and 0 cps (−10 and 0 ppm), that is, between the frequency of the reference, tetramethylsilane, and 600 cps less than the reference. In order to spread the spectrum over a convenient length of chart paper, the region from −1000 cps to −500 cps (−16.7 to −8.34 ppm) is commonly run in one scan, and the region from −500 to 0 cps (−8.34 to 0 ppm) in a separate scan. Throughout this book, the lowest trace will be the −1000 to −500 cps scan. It will be seen in Figure 2-2 that the region from −1000 to −500 cps contains no signals.

There are three characteristics of the tetramethylsilane signal which should be noted. (1) The right-hand side, the higher-field side, of the peak displays an exponentially decaying oscillation called ringing. An exaggerated example of ringing is shown on the left-hand side of Figure 2-2. (2) The tetramethylsilane signal occurs at 0 cps in the spectrum. Actually this signal is adjusted to 0 cps in the spectrum by the operator of the instrument. If this signal is not at 0, the entire spectrum should be corrected accordingly. (3) The tetramethylsilane signal is a symmetrical peak. An improperly operating instrument can produce an unsymmetrical signal.

There are two other ways of expressing the magnetic field strength scale other than in cycles per second. If the magnetic field strength scale in cycles per second is divided by the frequency of the signal used, which is 60 Mcps on the A-60, a dimensionless scale is obtained which is commonly referred to as the delta (δ) scale. The delta units are expressed in parts per million (ppm) and are also printed on the usual chart paper. A third way of indicating the magnetic field strength is by use of the tau (τ) scale [76], which is obtained by substracting the delta scale from 10, a number assigned to TMS. A signal which occurs at 60 cps downfield from TMS at 60 Mcps would thus be said to be 1 ppm downfield, or at $\tau = 9$. Chart paper is available which is calibrated with all three scales. The symbol commonly used for the cycles per second scale is ν. The difference in two positions in cycles per second is designated as $\Delta\nu$.

There are strong proponents of each of the three scales used for expressing magnetic field strength. The arguments for each of the scales, which will be more readily understood after a consideration of the remainder of this chapter, are summarized in

Table 2-3. In this book, the values will be reported in cycles per second as shifts downfield from TMS at 60 Mcps. Since all of the shifts which will be reported are to lower fields, the negative sign will be omitted. Corresponding values on the delta scale (ppm) will also be given. The tau values can be readily obtained by subtracting the delta values from ten. All three scales appear on most of the spectra.

It should be noted that the sign of the chemical shift is sometimes reported in the literature as positive and sometimes as negative. This causes little difficulty, however, if it is remembered that nearly all proton signals occur at lower magnetic fields than TMS.

It turns out that the separation of two sharp, single peaks, expressed in cycles per second, in the NMR spectrum is proportional to the frequency used to make the measurement (see Figure 2-3). This means that if a proton absorbed at 600 cps with reference to TMS at 60 Mcps, this same proton would absorb at $\frac{40}{60} \cdot 600$ or

TABLE 2-3
Arguments for Each of the Three Scales Used in Expressing Magnetic Field Strength

Scale	Arguments for the use of the scale
ν (cps)	(1) All instruments are calibrated in cps.
	(2) Coupling constants must be reported in cps.
	(3) Unless the actual chemical shifts are known, positions of peaks should be reported in cps.
	(4) Most laboratories have only one frequency available, and thus they are not greatly concerned with the frequency-dependence of chemical shifts.
	(5) Double resonance experiments are set up in terms of cps.
δ (ppm) (ν divided by the frequency used)	True chemical shifts expressed in ppm are independent of the frequency used.
τ (ppm) (for TMS: $10-\delta$)	(1) True chemical shifts expressed in ppm are independent of the frequency used.
	(2) The scale numbers increase from left to right.
	(3) There are fewer negative numbers.

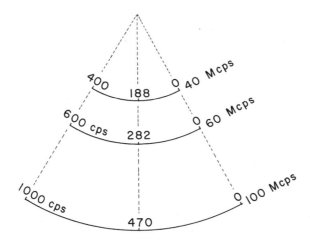

Fig. 2-3. Separation of single, sharp peaks as a function of
the frequency used.

400 cps with reference to TMS at 40 Mcps. At 100 Mcps, the separation of single sharp peaks would be $^{100}\!/_{60}$ times the separation observed at 60 Mcps.

Data obtained under different conditions must first be converted to the frequency of interest and then corrected for the particular reference used.   This conversion scheme is shown in Table 2-4.

TABLE 2-4

Conversion   of   Literature   Values   to
Chemical Shifts in Cycles per Second at
60 Mcps Using TMS as Internal Reference

I.	Convert to cps at 60 Mcps:	
	A.	cps · (60 divided by the frequency used)
	B.	δ (ppm) · 60
	C.	$(10-\tau)$ · 60
II.	Correct for reference used:	
	cyclohexane	86 cps
	water	282 cps
	benzene (internal)	438 cps
	benzene (external)	384 cps

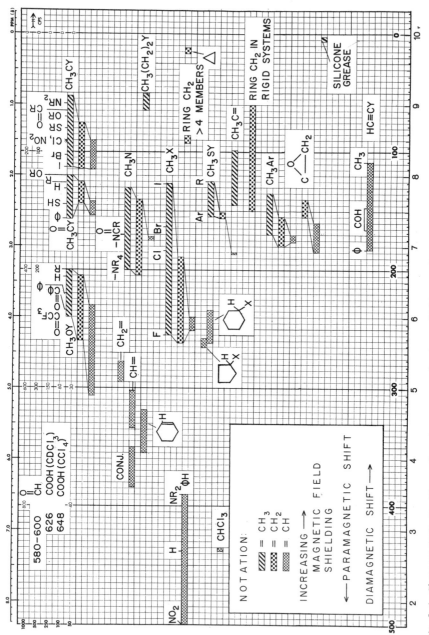

Fig. 2-4. Characteristic positions of various proton signals at 60 Mcps with reference to internal TMS. Unless otherwise noted, the positions are those of aliphatic protons. The data are from the text by Jackman (Chapter 4) [2] and the catalogs by Bhacca, Johnson, and Shoolery and Bhacca, Hollis, Johnson, and Pier [7,8]. A larger version of this chart will be found in a pocket on the inside back cover.

For example, Shoolery and Rogers reported [20] that at 40 Mcps the the C-6 proton in a $\Delta^{5,6}$- steroid absorbs at 42 cps upfield from external benzene (benzene contained in a sealed capillary rather than dissolved in the solution). If this measurement has been performed at 60 Mcps, the separation of the C-6 proton signal and the external benzene signal would have been 42 cps $\cdot$ $^{60}/_{40}$ or 63 cps. At 60 Mcps, the protons of external benzene absorb at −384 cps with reference to TMS. This means that the C-6 proton should absorb at −384 + 63 or −321 cps using a signal of 60 Mcps and TMS as an internal reference.

Figure 2-4 shows the positions at which many protons absorb at 60 Mcps with reference to TMS. Unless otherwise noted, these positions are given for aliphatic compounds. A few commonly employed terms are given on the left-hand side of this chart. It is conventional to display the magnetic field strength increasing from left to right. A proton which absorbs near TMS is said to be more shielded than a proton which absorbs to the left. A shift from left to right is spoken of as a diamagnetic shift and a shift from right to left is said to be a paramagnetic shift.

Only a few protons absorb at magnetic fields higher than TMS. Among these protons are those attached to heavy metals. The protons in cyclopropane derivatives and silicone stopcock grease absorb at a slightly lower field than TMS.

There are three generalizations which can be made concerning the position at which various protons absorb in the spectrum. (1) Increasing the electronegativity of a substituent causes a shift of the proton signal to the left (a downfield or paramagnetic shift). Thus protons in methyl groups attached to saturated carbon atoms appear between 40 and 110 cps (0.67 and 1.83 ppm), protons in methyl groups attached to nitrogen atoms appear between 130 and 200 cps (2.16 and 3.33 ppm), and protons in methyl groups attached to oxygen atoms absorb between 200 and 240 cps (3.33 and 4.00 ppm). (2) In general, the proton in a methine group ($-\overset{|}{\underset{|}{C}}H$) absorbs to the left of the corresponding protons in a methylene group ($\overset{|}{\underset{|}{C}}H_2$), which, in turn, absorb to the left of the corresponding protons in a methyl group ($-CH_3$). (3) The third generalization is that the position of a proton signal is dependent, in part, on the spatial relationship of the proton to certain functional groups (see Figure 2-5). For example, the signal due to a proton which is placed in the

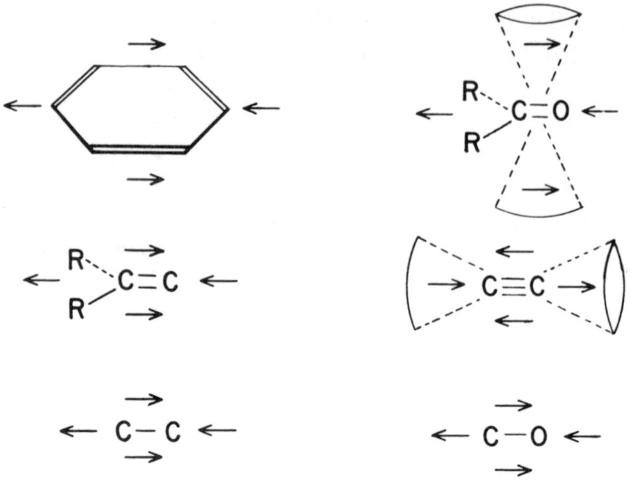

Fig. 2-5. Direction of shift of proton signal due to the spatial
relationship of the proton with various functional groups

plane of a phenyl group is shifted to the left, to lower field, and
the signal due to a proton placed in the plane of a carbonyl or olefinic group or in line with a carbon–carbon or carbon–oxygen single
bond is shifted to the left, while the signal due to a proton placed
above one of these groups is shifted to the right. The acetylenic
group has the opposite effect in that the signal due to a proton which
is placed in line with the acetylenic group is displaced to the right,
while the signal of a proton placed above the acetylenic group is
shifted to the left. These effects explain, in part, the very low
fields at which aldehydic protons (580 to 600 cps; 9.67 to 10 ppm)
and aromatic protons (390 to 500 cps; 6.50 to 8.33 ppm) absorb.
The apparently anomalous position of the acetylenic proton signal
(about 140 cps; 2.33 ppm), which is between aliphatic and olefinic
proton signals, is also explained in this fashion.

Two of these spatial effects have been dealt with in a semiquan-
titative fashion. Johnson and Bovey [21] have plotted the change in
field which would be expected to result from the change in the position of a proton as a function of the distance (A) above the center of
the plane of the ring and distance (B) from the center of the ring in
the plane of the ring (see Figure 2-6). This particular correlation
is especially useful when one is dealing with two isomers in which

there is a variation in the spatial relationship of a distinctive proton and a phenyl group. Jackman [22] has shown that for rigid a,a'-unsubstituted cyclohexane derivatives the calculated difference in frequency between an equatorial and an axial proton (about 24 cps; 0.40 ppm) is in good agreement with the observed difference (see Figure 2-7).

The position at which protons in aliphatic methylene groups absorb can be predicted by use of Shoolery's additive constants [23,24] (Table 2-5).   For example, the methylene protons in methylene

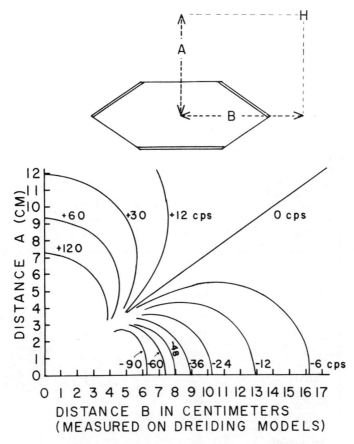

Fig. 2-6. Effect on the position of absorption due to the spatial relationship of a proton with a phenyl group. After Johnson and Bovey [21]. The distances are measured on Dreiding stereomodels (1 cm = 0.4 A).

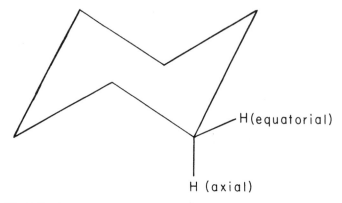

Fig. 2-7. The axial proton in an α, α'-unsubstituted, rigid cyclohexane system often appears at higher fields (to the right) than the
corresponding equatorial proton. This results from the spatial effect
of the carbon—carbon bonds.

TABLE 2-5

Shoolery's Additive Constants [23,24]

The position in cps of the signal due to an aliphatic methylene
group in XCH_2Y in CCl_4 is equal to the sum of the constants
(cps) for the two substituents plus 14 cps. The position in ppm
is equal to the constants (ppm) plus 0.233 ppm. The constants
may be used, but with much less success, for methine proton
positions. Data is for 60 Mcps with TMS as internal reference.

Group	σ (cps)	σ (ppm)
Cl	152	2.53
Br	140	2.33
I	109	1.82
C_6H_6	110	1.83
NR^1R^2	94	1.57
OAlkyl	142	2.36
SR	98	1.64
$CR=O$	102	1.70
$CR^1=CR^2R^3$	79	1.32
$C\equiv CH$	86	1.44
$C\equiv N$	102	1.70
CH_3	28	0.47

chloride would be predicted to absorb near 152 + 152 + 14) or 318 cps, 5.29 ppm, which is in good agreement with the found position of 316 cps (5.26 ppm). The constants in Table 2-5 can also be used to predict the position of absorption of methine ($-\overset{|}{\underset{|}{C}}H$) protons, but the agreement with the found value is not usually as good as with the methylene proton positions.

A rapidly increasing number of tables of additive constants can be found in the literature. R. F. Zürcher [25] has given the constants necessary to predict the positions of both the C-19 methyl protons and the C-18 methyl protons in many steroids. Using these constants, the C-19 and C-18 methyl protons of the steroid in Figure 2-8 would be predicted to absorb at 79.0 cps (1.318 ppm) and 42.5 cps (0.709 ppm), respectively. The observed values are 79.8 cps

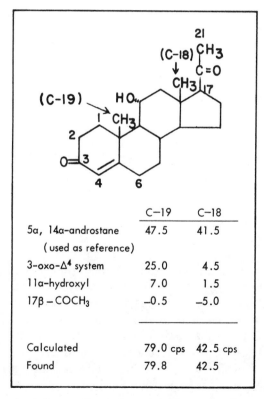

	C–19	C–18
5α, 14α-androstane	47.5	41.5
(used as reference)		
3-oxo-Δ⁴ system	25.0	4.5
11α-hydroxyl	7.0	1.5
17β – COCH₃	–0.5	–5.0
Calculated	79.0 cps	42.5 cps
Found	79.8	42.5

Fig. 2-8. Example of the use of Zürcher's additive constants [25] for the C-19 and C-18 methyl protons of steroids. See also Ref. 19.

and 42.5 cps, which are in good agreement with the calculated positions.

The characteristic absorption position of only one proton attached to oxygen (—COOH) is given in the correlation chart (Figure 2-4). The reason for this is that the field strength at which a proton absorbs is very sensitive to hydrogen bonding, which in turn is dependent upon the compound, solvent, temperature, and concentration. As a consequence, the exact position at which a proton attached to oxygen absorbs is usually difficult to predict. Most carboxylic acids exist as dimers, thus making the degree of hydrogen bonding more constant and the position (about 626 cps; 10.43 ppm) of the proton attached to oxygen more predictable. One other difficulty which arises is the result of the effect of exchange of protons between different environments. If, for example, a molecule contains two hydroxyl groups which are in quite different molecular environments, the two different hydroxyl protons may appear in different parts of the spectrum (Figure 2-9a). If these two different protons begin to change places with each other, the two signals become broader (Figure 2-9b). When the rate of exchange in times per second approaches the original separation of the peaks in cycles/sec, the two broad peaks coalesce into a single broad peak (Figure 2-9c) [26]. At even faster exchange rates, the broad peak becomes quite sharp (Figure 2-9d).

The signal due to a proton attached to sulfur is influenced by the same factors as the signal due to a proton attached to oxygen.

The prediction of positions of signals due to protons attached to nitrogen is complicated not only by the effects of hydrogen bonding and exchange, but also by properties of the nitrogen nucleus which can cause a proton attached to it to appear as a sharp single peak, a broad single peak, or three broad peaks. The protons on the nitrogen in most aliphatic and aromatic primary amines appear as a single peak, pyrrole and amide protons appear as broad peaks, and amines in strong acid solutions may appear as three broad peaks.

The ambiguities in spectra caused by protons attached to oxygen, nitrogen, and sulfur are easily removed. If, after the spectrum has been determined in deuterochloroform, a few drops of heavy water (D_2O) is added and this mixture is shaken for several minutes, all of the protons attached to oxygen, nitrogen, and sulfur will be replaced by deuterium, which does not show up in the spectrum [72,73]. This exchange process is illustrated in the spectrum of

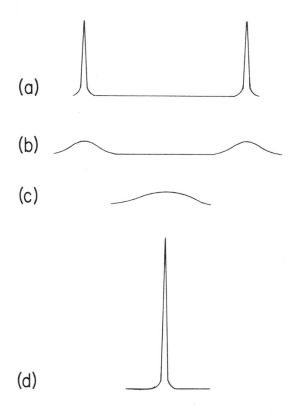

Fig. 2-9. Effect of the increase in exchange rate of protons between two different environments. In (a) the exchange is very slow. As the rate increases, the peaks broaden as in (b). The two broad bands finally coalesce into one broad band (c). At high exchange rates (d), a single sharp peak is observed.

7-acetylpodocarpic acid (Figure 2-10). In the lower curve (scanned from 1000 to 0 cps) the absorption band at 710 cps (11.8 ppm) is due to the phenolic proton, while the band at 640 cps (10.7 ppm) is due to the carboxylic acid proton. The upper absorption curve was obtained after the deuterochloroform solution had been shaken with heavy water. It will be seen that the carboxylic acid proton signal has been removed completely, and only a small amount of the phenolic proton signal remains. Frequently a trace of water shows up near 282 cps (4.70 ppm) in the exchange spectrum.

In Figure 2-10, the TMS band is symmetrical and shows the ex-

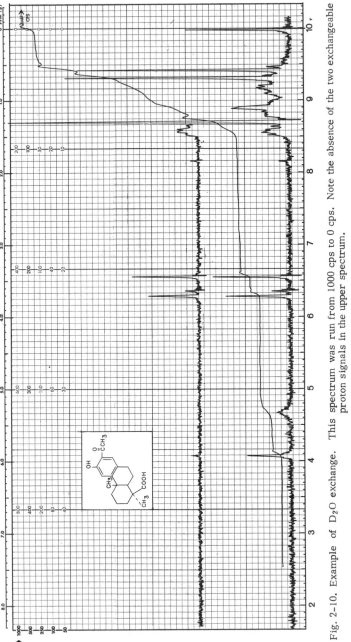

Fig. 2-10. Example of D_2O exchange. This spectrum was run from 1000 cps to 0 cps. Note the absence of the two exchangeable proton signals in the upper spectrum.

pected ringing, but is not at 0 cps. The entire spectrum should be corrected by subtracting 3 cps from each peak position. The small peak near 436 cps is due to a small amount of chloroform in the deuterochloroform. The very small peak near 218 cps is probably due to a trace of the corresponding methyl ester. The two tertiary methyl groups show up as strong peaks at 67 cps (1.12 ppm) and 80 cps (1.33 ppm). The protons in the acetyl group appear at 154 cps (2.57 ppm). Just to the left of the acetyl methyl signal is a complex band due to the protons in the methylene group adjacent to the phenyl group. The two aromatic protons appear at low fields

TABLE 2-6

Summary of Methods of Predicting Chemical Shifts and the General Rules Governing These Shifts

Chemical shifts can be predicted by the use of:

I. Correlation charts such as Figure 2-4 and those given by Stothers [15] and by Mohacsi [67].

II. Shoolery's additive constants (Table 2-5).

III. Tables of additive constants such as those of Zürcher [25].

IV. Reference to catalogs of spectra [7,8] or published spectra [9].

V. Closely related compounds.

In general, a shift to the left (to lower field) is caused by:

I. Increasing the electronegativity (electron-withdrawing effect) of a substituent.

II. Going from methyl ($-CH_3$) to methylene ($-CH_2-$) to methine ($-\overset{|}{\underset{|}{C}}H$).

III. Placing the proton (1) in the plane of a phenyl ring, carbonyl group, or olefinic bond, or (2) in line with a carbon–carbon or carbon–oxygen bond, or (3) above an acetylenic bond. [The signal of a proton placed (1) above the plane of a phenyl ring, carbonyl group, or olefinic bond, or (2) above a carbon–carbon or carbon–oxygen single bond, or (3) in line with an acetylenic bond is shifted to the right.]

IV. Increasing the hydrogen bonding to a proton.

The positions of signals due to protons attached to oxygen, nitrogen, and sulfur are difficult to predict because of variations in the degree of hydrogen bonding. These signals are easily identified by the D_2O exchange technique.

(410 and 443 cps; 6.83 and 7.38 ppm), in part because they are in the plane of the phenyl ring. The remaining ring protons give rise to the broad background absorption between 50 and 154 cps (0.834 and 2.57 ppm). The very low field at which the phenolic proton appears results in part from the electron-withdrawing effect of the phenyl, the hydrogen bonding of the proton to the carbonyl group, and the spatial effect of the carbonyl group.

Facts which should be remembered about chemical shifts are summarized in Table 2-6.

Not only can the bands due to exchangeable protons be identified by the D_2O exchange technique, but the number of the protons can be determined by use of the integration curve.

It turns out that, under the proper operating conditions, the area under an absorption band is proportional to the number of protons responsible for the absorption. By a comparison of the areas under different absorption bands, the relative number of protons represented by each band can be determined. The Varian A-60 instrument is provided with an electronic integrator. This integrator provides a curve which consists of a series of plateaus (see Figure 2-11). The integration is conventionally run from lower to higher fields. The height of a particular plateau from the starting point of the integration is thus proportional to the number of protons up to that point in the spectrum. The difference in height between adjacent plateaus is proportional to the number of protons causing the increase in elevation. Occasionally it is desirable to reset the integrator to the base line at one or more places in the spectrum. The elevations can be measured in any units. The squares on the chart paper are frequently used. In Figure 2-11 the total elevation, neglecting the increase due to the impurity, was 14 units.

There are a number of ways of dealing with the integration curve. If a structure can be proposed for the compound, the total increase in elevation can be assigned to the number of protons in the proposed structure. In Figure 2-11, for example, if the structure is assumed to have seven protons, each unit represents $^7/_{14}$ or 0.5 proton. This proportionality constant can then be applied to each portion of the spectrum. The band on the extreme right caused an elevation of two units and thus represents $2 \cdot 0.5$ or one proton. The advantage of dividing the number of protons by the total increase in elevation rather than dividing the elevation by the number of protons is that when a slide rule is used the slide is then set up

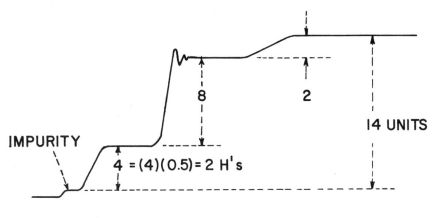

7 H's / 14 UNITS = 0.5 H / UNIT

Fig. 2-11. Idealized integration curve.

for multiplying each increase in elevation by the proportionality constant. Another way of using the integration curve is to assign a particular band to a given number of protons. If the molecule were suspected of having a methoxy group, the methoxy peak, which would appear in the 200-cps (3.33 ppm) region, could be assigned to three protons. The proportionality constant thus obtained could then be used throughout the spectrum. The band having the smallest area could be assigned to one proton and the resulting factor could then be applied to the other bands.

It is to be emphasized that the integration curve reveals only the ratio of the number of protons causing each band. The methods cited above yield absolute numbers of protons only when the correct assumption is made about the total number of protons in the molecule or group. If in Figure 2-11 the structure had been assumed to contain 14 rather than seven protons, the two-unit area would represent two rather than one proton. A molecular weight determination by a method such as mass spectroscopy or isothermal distillation is quite helpful in utilizing the integration curve.

Sharp, strong peaks are followed by ringing and overshooting of the recorder pen as indicated in the eight-unit peak in Figure 2-11. The height is measured after these two effects have been averaged out.

There is no completely satisfactory way of dealing with the separate integration of overlapping bands. Some aid can be obtained by the use of closely related model compounds which lack one of the bands. An expansion of the absorption curve is often helpful. A change of solvent or the use of a higher frequency may cause the bands to separate. These techniques are discussed in Chapter 5.

An actual integration curve is shown in Figure 2-12. The general features of the spectrum will be examined first. The symmetrical TMS peak is almost at 0 cps and shows good ringing. The two tertiary methyl groups appear at 64 cps (10.7 ppm) and 76 cps (1.27 ppm). The protons in the two methoxy groups occur at 220 cps (3.67 ppm) and 235 cps (3.92 ppm). The broad bands centered at 370 cps (6.16 ppm) and 460 cps (7.67 ppm) are due to the protons in the primary amide group. The carbon–nitrogen bond has enough double-bond character to hold the two protons in relatively fixed cis and trans relationships to the carbonyl group. Thus the primary amide behaves as if the structure were

$$
\begin{array}{c}
O \\
\diagdown \\
C = N \\
\diagup \qquad \diagdown
\end{array}
\begin{array}{c}
H \\
\diagup \\
\\
H
\end{array}
$$

in which the two amide protons have different environments. The appearance of two broad signals for primary amides is fairly common. The two sharp signals at 410 cps (6.84 ppm) and 472 cps (7.87 ppm) are due to the phenyl protons. A trace of chloroform appears at 436 cps (7.27 ppm).

There are 27 protons in the structure proposed for the compound in Figure 2-12. The total integration amounts to 38.5 units, of which 0.3 unit is due to chloroform as an impurity. Thus each unit represents $^{27}/_{38.2}$ or 0.706 proton. The number of protons represented by each band is then equal to 0.706 times the units of integration caused by that band. For example, the two unresolved bands in the 450 to 476 cps region caused an increased elevation of 2.8 units and hence represent $2.8 \cdot 0.706$ or 1.98 protons. One of these protons is on the aromatic ring and the other proton is on the amide nitrogen. The number of protons represented by some of the other bands is also indicated in Figure 2-12.

If the compound used for Figure 2-12 had been a dimer instead of a monomer, all of the calculated values based on the monomeric structure would have been half the actual values.

The estimates of the number of protons in individual groups obtained in this manner are usually accurate within 10%. It must be remembered that the integration is in arbitrary units, and the actual total elevation is under the control of the operator. The operator

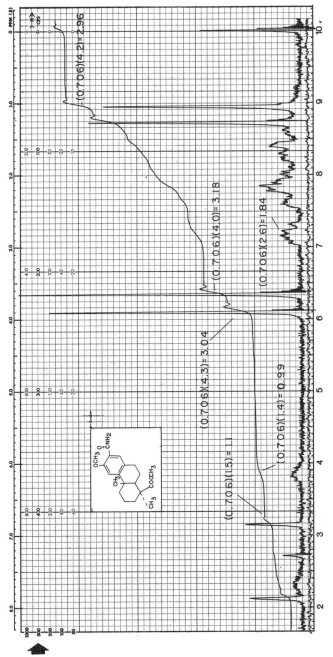

Fig. 2-12. Example of an actual integration curve.

of the instrument also determines the vertical starting point of the integration. Care must be exercised to account for impurities in the solvent employed and to ignore the integration of the TMS signal.

The spectrum of a typical steroid, 17α-methyltestosterone, is shown in Figure 2-13. The TMS signal appears normal. A small amount of chloroform appears at 437 cps (7.28 ppm). The calculated positions of the C-18 and C-19 methyl protons, using Zürcher's data [25], are 55.0 cps (0.917 ppm) and 73.0 cps (1.22 ppm). There are signals at these two positions, but the peak at 73.0 cps is of much greater intensity. The C-17 methyl group must therefore be accidentally equivalent to the C-19 methyl group, and these two groups appear as a single peak at 73.0 cps. The protons next to the carbonyl group and the double bond (C-2 and C-6 methylene protons) have signals in the 130-155 cps (2.16-2.58 ppm) region. The remaining ring protons on saturated carbon atoms give rise to broad background absorption in the 40-130 cps (0.67-2.16 ppm) region. The single C-4 olefinic proton appears at 344 cps (5.74 ppm).

Part of the absorption peak near 100 cps (1.67 ppm) in Figure 2-13 can be assigned to the hydroxyl proton on the basis of the D_2O exchange spectrum (not shown).

When the carbon–carbon double bond in 17α-methyltestosterone is moved from the C-4 position to C-1 position, the expected change in positions of the angular methyl groups occurs (see Figure 2-14). The calculated positions are 53.5 cps (0.89 ppm) for the C-18 methyl protons and 63.0 cps (1.05 ppm) for the C-19 methyl protons. The C-17 methyl protons appear as a separate peak at 73 cps (1.22 ppm). Only the two protons at C-4 show up in the 125 to 145 cps (2.08 to 2.42 ppm) region. The remaining protons on saturated carbons appear in the broad band between 40 and 125 cps (0.67 and 2.08 ppm). The sharp peak at 88 cps (1.47 ppm) is due to the C-17 hydroxyl proton.

A very curious series of peaks appears in the olefinic region of Figure 2-14. The expected chloroform peak occurs at 437 cps (7.28 ppm), but, rather than two peaks for the two olefinic protons, there is a pair of doublets. The integration curve shows that each doublet represents one proton. Actually, the doublet at the left near 430 cps (7.17 ppm) is due to the C-1 proton, while the doublet near 350 cps (5.84 ppm) is due to the C-2 proton. The splitting of the two bands of the two olefinic protons is a consequence of what is called spin–spin coupling. Each proton behaves like a small bar magnet which can be oriented in one of two ways. The orientation in one direction (α) causes the signal due to the other proton to be shifted to

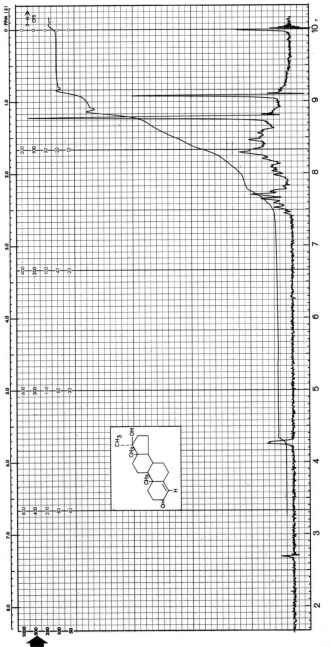

Fig. 2-13. Spectrum of 17α-methyltestosterone.

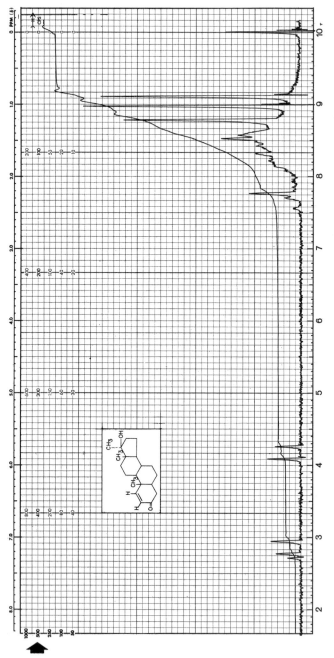

Fig. 2-14. Spectrum of 17α-methyl-17β-hydroxy-3-oxo-1-androstene.

the left, while orientation in the other direction (β) causes the signal to be shifted to the right. The result is the appearance of two signals, one on each side of the original position. The area under the two signals represents one proton. A more detailed representation of this region of the spectrum is shown in Figure 2-15. Each pair of peaks is separated by exactly the same magnetic field strength, 10 cps. This separation is called the coupling constant and is represented by the letter J. The value of J is independent of the frequency used to make the measurement. That is, the separation of the members of the doublets at 30 Mcps would be exactly the same (10 cps) as that found at 60 Mcps. The value of J is thus always expressed in cycles per second and never in delta or tau units. The larger the value of J, the stronger the coupling is said to be. The coupling constant can have either a positive or negative sign.

The position at which the C-1 and C-2 protons would be expected to absorb in the absence of spin–spin interaction can be easily calculated using the relationships given in Chapter 4. These positions are indicated in Figure 2-15. The ratio of the calculated separation (77.3 cps) of the peaks in the absence of spin–spin interaction to the coupling constant (10 cps) is an important quantity. In this particular case, this ratio ($\Delta\nu/J$) is about $^{77}/_{10}$ or 7.7.

The olefinic proton region shown in Figure 2-15 is approximately centered on a scan 250 cps wide. The TMS signal is offset 254 cps to the right. This means that the actual positions of the peaks are found by adding 254 cps to the numbers found on the 0 to 250 cps scale. Thus the signal appearing at the highest field (farthest to the right) is actually at 91 (position on the 0 to 250 cps scale) plus 254 cps (offset) or 345 cps. Expansions of this type, which are easily made, are often very useful in the detailed examination of selected regions of the spectrum.

The separation of the peaks in the absence of spin-spin interaction is directly proportional to the frequency used to make the measurement. The separation of the spin–spin doublets (J) is independent of the frequency used. As a consequence, if the frequency were decreased, the pattern would change as shown in Figure 2-16.

A secondary effect of decreasing the ratio of $\Delta\nu$ to J, which can also be seen in Figure 2-16, is that the intensities of the inner lines increase at the expense of the outer lines.

Under what conditions do protons interact to cause mutual splitting? Protons on the same atom and on adjacent atoms usually couple with each other, but protons on nonadjacent atoms usually do not couple strongly unless there are intervening multiple bonds.

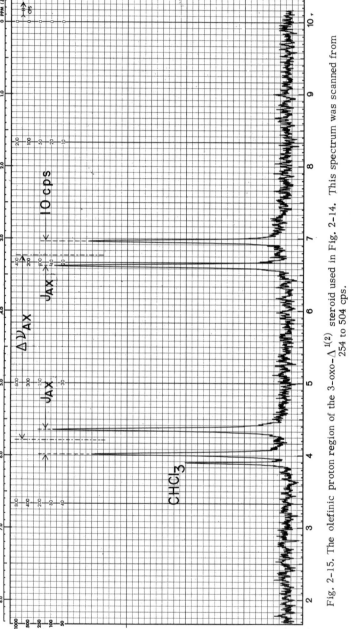

Fig. 2-15. The olefinic proton region of the 3-oxo-$\Delta^{1(2)}$ steroid used in Fig. 2-14. This spectrum was scanned from 254 to 504 cps.

Fig. 2-16. Representation of the effect of decreasing the frequency used on the spin–spin pattern of Fig. 2-15. The spacing between the doublets (J) remains constant while the doublets are brought closer together.   The intensities of the inner peaks grow at the expense of the outer peaks.

Although the spatial features of the system are important, spin–spin coupling is primarily through bonding electrons. Tables 2-7a, b, and c give both the usual range of J for various arrangements of atoms and also the more typical values.   It will be seen that the coupling constant, which can be either positive or negative, seldom exceeds ±20 cps.   The values for J given in the tables are a useful guide, but must be employed cautiously. Occasionally, coupling occurs through greater distances than indicated in Tables 2-7a, b, and c. Examples of long-range spin–spin coupling are given in Table 2-7d. The effects of substituents on the J's in the tables are, at the present time, not always predictable.

The coupling constant of protons on adjacent carbon atoms is a function, in part, of the angle between the plane $H_A C_1 C_2$ and $C_1 C_2 H_B$ [33] (see Figure 2-17). This angle, which is called the dihedral angle, is the angle seen between the bonds to the two protons when the molecule is viewed along the carbon–carbon bond.

In Table 2-7a it will be seen that in rigid cyclohexane systems the coupling of adjacent axial–axial protons is large (usually 8-12 cps), while the coupling of axial–equatorial and equatorial–equatorial protons is small (usually 3-4 cps). These relationships are a direct result of the effect of the dihedral angle on the coupling constant.

The coupling constant between protons on adjacent saturated carbon atoms is also influenced by factors other than the dihedral angle [34]. The coupling decreases (1) with an increase in the elec-

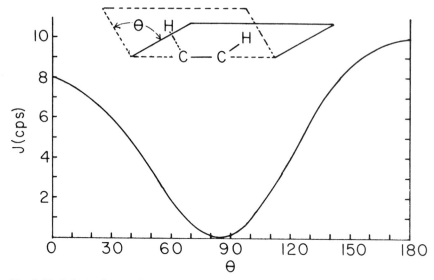

Fig. 2-17. Relationship of the coupling constant of protons on adjacent carbon atoms to the dihedral angle [33]. This angular relationship is only one of the factors which must be considered [34].

tronegativity of substituents, (2) with an increase in the carbon–carbon bond length, and (3) with an increase in the angles between the bonds to the protons and the carbon–carbon bond:

$$H_A \quad \theta \qquad \text{and} \qquad \theta \quad H_B$$
$$C{-}C \qquad \qquad C{-}C$$

The coupling constant between protons on the same carbon atom (geminal coupling constant) in a freely rotating system is approximately equal to −12.4 cps plus −1.9 cps for each adjacent double bond and (2)(−1.9), or −3.8 cps, for each adjacent triple bond [36b]. Thus in
$$R{-}\overset{\displaystyle \overset{O}{\|}}{C}{-}\overset{\displaystyle \overset{H_A}{|}}{\underset{\displaystyle \underset{H_B}{|}}{C}}{-}, \quad J_{AB} \text{ would be approximately } -12.4$$
+(−1.9) or −14.3 cps, while in
$$R{-}\overset{\displaystyle \overset{H_A}{|}}{\underset{\displaystyle \underset{H_B}{|}}{C}}{-}C \equiv N, \quad J_{AB} \text{ would be about}$$
−12.4 + (2)(−1.9) or −16.2 cps. In rigid systems the contribution of an adjacent double bond to the geminal coupling constant depends

TABLE 2-7a

Spin–Spin Coupling Constants for Protons on Saturated Systems

Most of the values in this chart and in Tables 2-7b and 2-7c are from the collections of (1) Jackman [27], (2) Pople, Schneider, and Bernstein [28], and (3) Hollis [29].   A survey of couplings between vicinal protons in cyclohexane systems is given by Huitric, Carr, Trager, and Nist [30]. The effect of a carbonyl on the coupling between geminal protons is discussed by Takahashi [31], and the effect of ring size on coupling in 1,3-dioxolans by Crabb and Cookson [32]. Many leading references are given by Bishop [13].

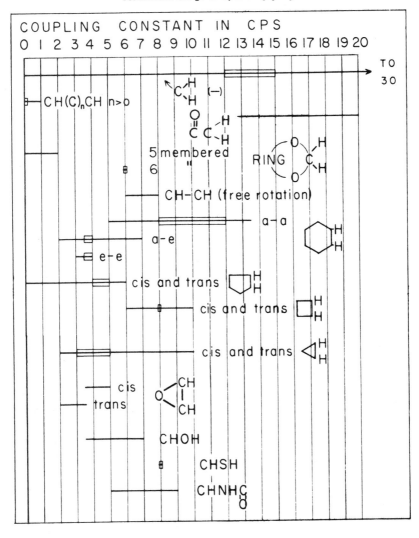

TABLE 2-7b

Spin–Spin Coupling Constants for Aldehydic Protons and Protons on Multiple Bonds

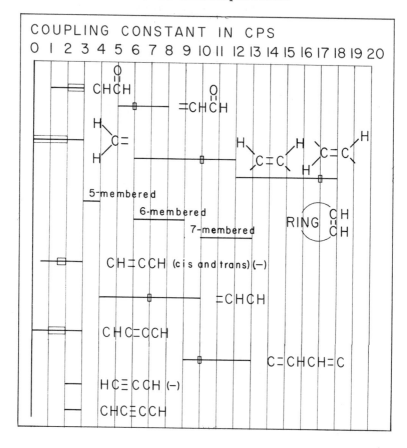

TABLE 2-7c

Spin–Spin Couplings for Protons on Aromatic Systems

A review of the NMR of heterocyclic compounds is given by R. F. M. White [16]. Other leading references are given by Sternhell [40a].

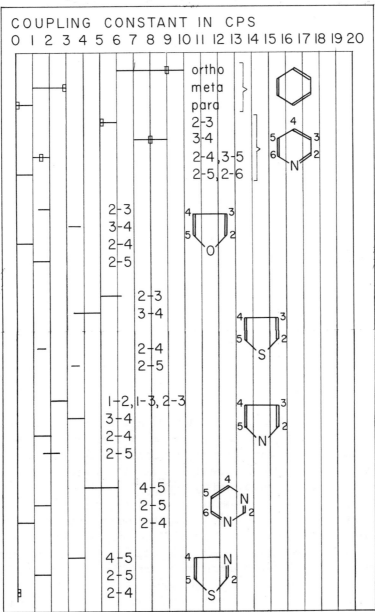

TABLE 2-7d

Some Examples of Long-Range Spin–Spin Interactions

The bonds linking the coupled protons in (b) form a figure W. This appears to be a general requirement for four-bond proton–proton coupling in saturated systems. See the review by Sternhell [40a].

(a) $CH_3C{\equiv}C{-}C{\equiv}C{-}C{\equiv}C{-}CH_2OH$ $J_{CH_3,CH_2} = 0.4$ cps [38]

(b) $J_{AB} \cong 7$ cps [39]

(c) $J_{AB} = 1.3$ cps [41]

(d) $J_{CH_3,H_B} = 0$ cps [42]

$J_{CH_3,H_A} = 0.65$ cps [42]

on the stereochemistry and varies from 0 to −5.7 cps. A theoretical curve (see Figure 2-18), derived several years ago, related the geminal coupling constant to the angle between the bonds to the two protons [35]. Although there is some dependence on this angle, this curve is no longer considered valid [36].

The coupling constant between protons on the same carbon atom is usually of opposite sign to that of the coupling constant between protons on adjacent carbon atoms [37]. It appears that the sign of the coupling constant alternates with each additional bond. That is, if the coupling of two protons on the same carbon is negative, the coupling of protons on adjacent atoms is positive, and so on. A change in the sign of the coupling constant in Figure 2-15 would cause no change in the appearance of the spin pattern. In systems involving more than two protons, however, a change in sign may alter the appearance of the pattern. These aspects are discussed in Chapter 4.

It was seen in Figure 2-15 that, when two protons are spin-coupled, each proton signal is split by the other proton into a doublet separated by J, the coupling constant. The number of peaks expected when more than two protons interact can easily be predicted if the following two simple conditions are met:

1. The difference in chemical shifts ($\Delta \nu$) between the groups of protons must be at least six times the coupling constant between the groups of protons. This condition is illustrated in Figure 2-19. In

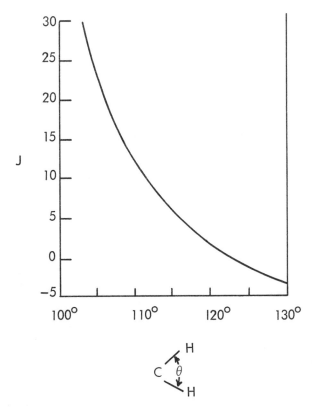

Fig. 2-18. A theoretical curve [35], no longer considered as valid [36], previously used for predicting the effect of a change in the angle between the bonds to protons attached to the same carbon atom.

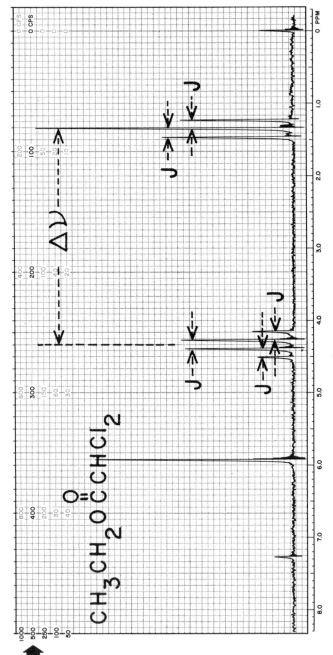

Fig. 2-19. One of the two conditions which must be met by systems before the simple multiplicity rules can be applied is that $\Delta\nu$ (the separation of the centers of the bands) be at least six times J. In this example, $\Delta\nu$ is 180 cps and J is 7 cps. Spectrum reproduced by permission of the copyright holder, Varian Associates.

aliphatic systems, where the coupling constant between protons on adjacent carbon atoms is typically 7 cps, the centers of the two bands must be separated by at least 42 cps.

2. Each proton in one group must be coupled equally to each and every proton in the second group. This second requirement, for the most part, restricts the groups to aliphatic systems in which three stable conformations are equally populated. Systems of the type $R\overset{\overset{\text{O}}{\|}}{C}OCH_2CH_2NR_2$ and $-OCH_2CH_3$ conform to this condition. Disubstituted phenyl groups such as

$$CH_3O-\underset{H\quad H}{\overset{H\quad H}{\langle\bigcirc\rangle}}-NO_2$$

do not conform to this second condition because each of the two equivalent protons *ortho* to the nitro group are not coupled equally to each of the two equivalent protons *ortho* to the methoxy group. From Table 2-7c it will be seen that protons *ortho* to each other on a phenyl ring are strongly coupled (9 cps), while protons *para* to each other are usually not coupled at all.

When the two conditions are met, the number of peaks in the bands is governed by the following four rules:

1. Equivalent protons give single, sharp peaks regardless of how strongly they are coupled to each other.
2. The multiplicity of peaks in the band arising from a group of equivalent protons is determined by the neighboring groups of protons.
3. The number of peaks in a band due to an equivalent group of protons is equal to the number of neighboring protons plus one. This will be refered to as the $N+1$ rule.
4. The peaks are symmetrically arranged about the chemical shifts of the groups and are separated from each other by the coupling constant.

These rules are easily understood by consideration of examples. The spectrum of diethyl succinate is shown in Figure 2-20. The center of the band due to the methyl groups (75 cps) is separated from the center of the band due to the adjacent methylene groups (250 cps) by 175 cps. The coupling constant between the two groups $(J_{CH_3-CH_2})$, which is equal to the separations between the components of the multiplets, is 7 cps. The ratio of the difference in

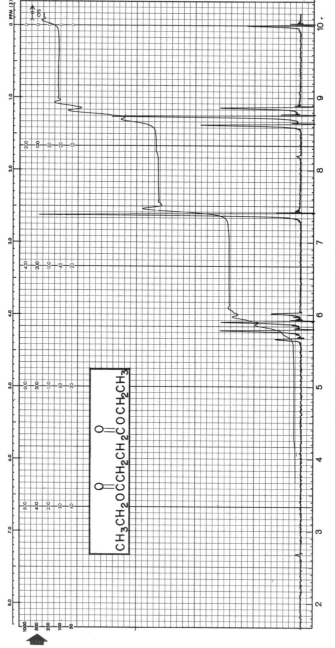

Fig. 2-20. Spectrum of diethyl succinate.

chemical shift ($\Delta\nu$) between the methyl and adjacent methylene groups to the coupling constant is equal to $^{175}/_7$ or 25 : 1. Rotation of the ethyl group about the carbon–carbon bond gives three equally stable, and hence equally populated, conformations. The rotation is rapid enough to average out the differences among the angularly dependent coupling constants. This then means that each proton in the methyl group is coupled equally to each and every proton in the adjacent methylene group. Both simple conditions are met by the two identical ethyl groups. The number of peaks expected for the methyl groups is then equal to the number of protons in the adjacent methylene groups plus one. Thus the methyl groups appear as 2 + 1 peaks or a triplet centered at 75 cps (1.25 ppm). The adjacent methylene protons appear as 3 + 1 or as a quartet centered at 250 cps (4.17 ppm). It is important to note that the multiplicity is always due to the number of neighboring protons, not to the number of protons in the group itself. The protons in the ethylene ($-CH_2-CH_2-$) group are all identical, and, even though they are coupled, equivalent protons cannot cause observable multiplicity. The equivalent ethylene protons can only appear as a single peak. This peak occurs at 157 cps (2.62 ppm).

The TMS signal in Figure 2-20 is symmetrical, shows some ringing, and occurs very close to 0 cps. The small peak near 108 cps is due to an impurity in the compound, while the peak at 439 cps is due to chloroform in the deuterochloroform.

The patterns produced by systems which conform to the two conditions ($\Delta\nu > 6J$ and all J's equal) are said to be first-order patterns. The systems themselves are called first-order systems.

Most of the pitfalls in the interpretation of spin patterns result from application of the first-order rules to systems which do not conform to one or both of the simple conditions. These problems are discussed in detail in Chapter 4.

One more characteristic can be stated about first-order patterns. The ratios of intensities of the multiplets are proportional to the coefficients of the terms in the expansion of $(R + 1)^N$ (see Table 2-8). Thus all first-order doublets have equally intense peaks, triplets have intensity ratios of 1 : 2 : 1, quartets 1 : 3 : 3 : 1, and so on. Often the outer peaks of higher multiplets are lost in the noise. It will be seen, however, that the intensities of the three strongest peaks in a quintet are in the ratio of 2 : 3 : 2 rather than the ratio expected for a triplet, 1 : 2 : 1.

The ratios of intensities of the peaks in Figure 2-20 is seen to closely approximate those predicted in Table 2-8 (1 : 2 : 1 and

TABLE 2-8

Approximate Ratio of Intensities of Peaks in First-Order Patterns

Number of peaks $(N + 1)$	Expansion of $(R + 1)^N$	Coefficients — ratios of intensities
2	$R + 1$	$1 : 1$
3	$R^2 + 2R + 1$	$1 : 2 : 1$
4	$R^3 + 3R^2 + 3R + 1$	$1 : 3 : 3 : 1$
5	$R^4 + 4R^3 + 6R^2 + 4R + 1$	$1 : 4 : 6 : 4 : 1$

$1 : 3 : 3 : 1$). The triplet has peaks in the ratio of $1.06 : 2 : 0.85$, and the quartet has peaks in the ratio of $0.96 : 3 : 2.98 : 1.08$. The observed distortion of the intensities is general. The peaks closer to the other part of the pattern are more intense than the corresponding peaks further from the other part of the pattern. More will be said about this distortion of intensities in Chapters 3 and 4.

All first-order patterns are characterized by: (1) symmetry about the midpoint of each part of the pattern, (2) equally spaced peaks, and (3) peaks having intensities in the approximate ratios of $1 : 1$, $1 : 2 : 1$, $1 : 3 : 3 : 1$, $1 : 4 : 6 : 4 : 1$, etc. The number of neighboring protons which causes the multiplet is one less than the number of observed peaks.

The facts which should be remembered about first-order systems and their patterns are summarized in Table 2-9.

A special notation [43] has been devised to facilitate the discussion of spin–spin coupling. Protons which absorb at about the same field are designated as A, B, C,.... Protons which absorb at positions quite different from the first set (A, B, C, ...) are designated by X, Y, ... The number of protons of the same type are indicated by subscripts. For example, methyl ethyl ether ($CH_3OCH_2CH_3$) has two essentially separate groups of protons, those in the methyl group and those in the ethyl group. From experience it is known that there is very little coupling between these two groups and that a large difference in the chemical shifts is expected for the methylene and methyl protons in the $-CH_2CH_3$ group. The methoxy protons would thus be said to constitute an A_3 system, while the ethyl group would be said to be an A_3X_2 system. If the signals of three groups of protons are far apart in the spectrum, the third group of protons is designated as M. For example, 1-nitropropane ($CH_3CH_2CH_2NO_2$) would be called an $A_3M_2X_2$ system.

It must be known from experience which protons are different, how different they are, and whether or not they are coupled. There is considerable arbitrariness in these designations. The protons which absorb to the right (higher fields) are generally, but not always, assigned the lower letters in the alphabet. There is no established demarcation at which an AB system becomes an AX system. In this book, a system will be designated as AX if the difference in chemical shift between A and X is at least six times the coupling constant between A and X. If $\Delta\nu_{AB}$ is less than six times J_{AB}, the system will be called an AB. The safest initial assumption to make about a system is that the protons are different, but not greatly different, and that they are all coupled.

One other point should be made in connection with this notation. Provision is not usually made in this scheme to indicate whether or not each proton in each group is equally coupled to each and every proton in the second group. For example, the aromatic portions in p-dimethylaminobenzaldehyde

might be called an A_2X_2 or an A_2B_2 system. A convention which will be followed in this book is to prime a system if each proton in one group is not coupled equally to each and every proton in the second group. Thus the aromatic protons above will be designated as an $A_2'X_2'$ system (the designation of this as an $A_2'X_2'$ rather than an $A_2'B_2'$ type results from an examination of the spectrum).

The following routine is suggested for the interpretation of spectra. First, the gross features of the spectrum should be noted, such as the range of frequencies recorded, the characteristics of the TMS signal, and the position of expected impurities, such as chloroform. An examination of the integration curve should then be made to determine the relative number of protons giving rise to each absorption band. The signals due to easily recognizable groups of protons known to be present should be identified. Among these signals should be those due to protons attached to oxygen and ni-

TABLE 2-9

Summary of Facts About First-Order Systems and Their Patterns

First-Order Conditions

I. $\Delta \nu > 6 J$.

II. Each proton in one group must be coupled equally to each and every proton in the second group.

First-Order Rules

I. Equivalent protons give only single peaks regardless of the strength of the coupling between the protons.

II. Multiplicity of a group of equivalent protons is caused by the number of neighboring protons, not by the number of equivalent protons.

III. The number of peaks is equal to the number of neighboring protons plus one.

IV. The peaks are symmetrical about the center of the chemical-shift position of the group and are separated by J.

V. The ratio of intensities of the peaks is $1:1, 1:2:1, 1:3:3:1, 1:4:6:4:1$, etc.

trogen, which can be detected by comparison of the spectrum before and after D_2O exchange. Identification of the other bands is then made using such aids as the chart of predicted positions (Figure 2-4), spectra of available model compounds [9], spectra in the Varian catalogs [7,8], and the chemical facts. Simple spin–spin patterns should be explained. Checks should be made on the agreement between the found and predicted coupling constants. A more detailed method of interpreting spectra is given in Chapter 6.

In the next chapter a closer examination of first-order patterns will be made. Unexpected nonequivalence of protons and proton couplings to other nuclei will also be discussed.

Chapter 3

First-Order Spin Patterns, Coupling of Protons with Other Nuclei, and Nonequivalence of Protons

To the organic chemist, the complications arising from spin–spin splittings are by far the most baffling of all of the aspects of NMR. Not only is he unfamiliar with the concept of spin–spin interaction and the quantum mechanical laws involved, but he is also unfamiliar with the language used to describe the systems and their interactions.

The results of the theoretical considerations of spin–spin interactions will be summarized in this and the following chapter. Both the pitfalls and the practical aids to the interpretation of spin patterns will be discussed. The notation introduced in Chapter 2 will be employed as frequently as possible to enable the student to develop confidence in his use of this system. A detailed approach to the interpretation of spin patterns utilizing all of these aspects will be given in Chapter 6.

If the centers of the absorption of two groups of nuclei are separated ($\Delta\nu$) by at least six times the coupling constant (J), and if each proton in one group is coupled equally to each and every proton in the second group, then the simple multiplicity rules apply. The resulting patterns are said to be "first-order." The spectrum of phenylethyl acetate, Figure 3-1, shows a first-order pattern produced by protons in adjacent methylene groups. The difference in the chemical shift between the two groups is 82 cps, and the coupling constant between the protons in the two groups is 7 cps. The ratio of $\Delta\nu$ to J is thus 11.7:1. The fast rotation about the carbon–carbon bond averages out any possible differences in the chemical shift between the protons attached to the same carbon atom. The three conformations about the carbon–carbon bond are almost equally populated. Thus both simple conditions are met. Each methylene group has two neighboring protons, and consequently each of the signals is split into 2 + 1 peaks. The peaks are sym-

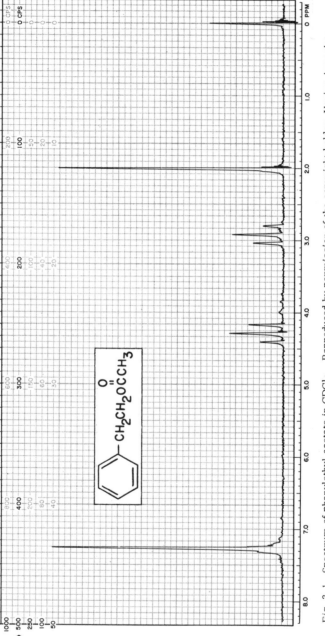

Fig. 3-1. Spectrum of phenylethyl acetate in CDCl₃. Reproduced by permission of the copyright holder, Varian Associates.

metrically arranged about the chemical shift positions of the two groups (175 and 257 cps) and are separated by the coupling constant, 7 cps. The ratio of intensities of the peaks, judged on the basis of the heights, is approximately $1:2:1$. The peaks of each triplet which are closest to the other part of the pattern are somewhat more intense than the peaks farthest from the other part of the pattern. This intensity distortion is equal in the two triplets. All first-order patterns show the same separation (J) between the peaks in each portion of the pattern (mutual coupling) and also show the same relative distortions of intensities.

The three protons in the methyl group are equivalent. There are no protons nearby with which they can couple. Even though they are coupled with each other $(J \approx -14 \text{ cps}?)$ equivalent protons cannot interact in such a way as to give observable multiplicity. The resulting single, sharp peak is at 120 cps (2.00 ppm). In the same manner, the very nearly equivalent aromatic protons give a single, fairly sharp peak at 435 cps (7.25 ppm).

It should be realized that the equivalence or nonequivalence of protons is often a matter of degree rather than an absolute condition. For example, at a very high frequency the phenyl protons in Figure 3-1 would probably appear as nonequivalent protons, since the slight differences in chemical shifts would be increased in proportion to the increase in frequency.

The protons in phenylethyl acetate constitute an A_5 (the phenyl protons), an A_2X_2 (the $-CH_2-CH_2-$group with the $-CH_2-O-$ protons as the X_2 part), and an A_3 system (the $-\overset{\overset{\text{O}}{\|}}{C}CH_3$ protons).

Occasionally, three different groups of protons are coupled, and the absorption bands are all separated by more than six times the corresponding coupling constants. For example, the three types of protons in nitropropane (Figure 3-2) are all separated by at least six times the corresponding coupling constants. In this molecule the coupling constant (about 7.0 cps) between the protons in the CH_3-CH_2- group is very nearly equal to the coupling constant between the $-CH_2-CH_2NO_2$ protons. The middle methylene protons thus appear as $3+2+1$ or six peaks which have the approximate ratio of intensities of $1:5:10:10:5:1$. The small peak at 96 cps (1.60 ppm) is separated by about 9 cps from the peaks of the middle methylene group. This peak must be due to an impurity. If it were part of the first-order multiplet, it would have to be separated by about 7-7.5 cps from the closest peak.

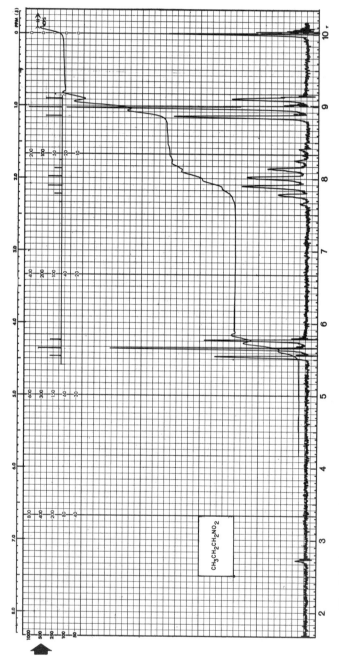

Fig. 3-2. Spectrum of 1-nitropropane in CDCl$_3$. The upper diagram is the pattern predicted by first-order rules.

The middle methylene proton signals are broadened somewhat by the small difference in coupling constants with the other two groups.   If this difference in coupling constants were increased, these peaks would each be split into two peaks, giving rise to a total of 12 peaks.   In other words, the methyl protons would split the middle methylene proton signal into $3+1$ or four peaks, and each of these four peaks would be split into $2+1$ or 3 peaks by the protons in the $-CH_2-NO_2$ group.   The multiplicity expected, then, for first-order splitting of one group of protons (M_P) by two other groups of protons (A_N and X_R) with which the M_P protons are unequally coupled is given by $(N+1)(R+1)$.   There are thus $N \cdot R$ more peaks when the coupling constants are not equal.   Both the three methyl protons and the two methylene protons next to the nitro group appear as triplets since they are each split by the two protons in the middle methylene group.

A graphical method is commonly used to visualize these first-order splittings.   A single line is first placed on graph paper to represent each group of equivalent protons (see Figure 3-3).   The spacings between the lines are made equal to the differences in

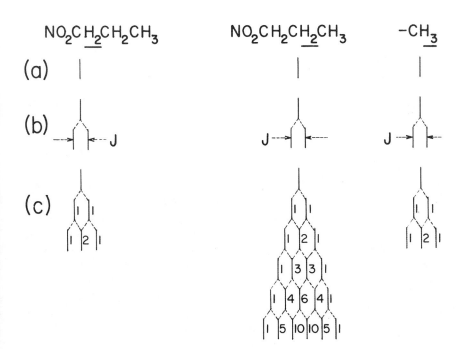

Fig. 3-3. Construction of a first-order splitting diagram for 1-nitropropane.

chemical shifts. The three groups of protons in 1-nitropropane would thus be represented as shown in Figure 3-3a. Next, each line representing each group of protons is split into a doublet, separated by J, if this group is coupled with one or more neighboring protons, as shown in Figure 3-3b. The same process is now repeated for each additional proton with which the group is coupled.

The completed diagram is shown in Figure 3-3c. The bottom row shows the number of peaks predicted for each group. It will be recalled that in first-order splittings the ratio of intensities of the doublets is $1:1$. If the intensities where lines coincide are added, the binomial ratios can be deduced from Figure 3-3c. Thus the center line of each triplet results from two equal coincident lines. In the same way, the additions of the intensities of the other converging lines will give the predicted relative intensity for the resulting peak. The intensities have been added in this way in Figure 3-3c. It will be seen that the ratios are the same simple ratios indicated earlier for first-order multiplets.

Had $J_{CH_3CH_2}$ been different from $J_{-CH_2CH_2NO_2}$, the $\underline{CH_3}$ protons and the $-\underline{CH_2}NO_2$ protons would still have been triplets, but the $CH_3\underline{CH_2}CH_2NO_2$ protons would have consisted of a maximum of $(3 + 1)(2 + 1)$ or 12 peaks. The extra peaks would arise because fewer of the lines would converge. There could, of course, be special intermediate cases for systems in which the ratio of $J_{CH_2CH_2NO_2}$ to $J_{CH_3CH_2}$ was $1:2$ or $2:1$. This would reduce the number of peaks. The number of peaks for this first-order system would always be between that of the equal coupling case of $3 + 2 + 1$, or 6, and the maximum for the unequally coupled case, $(3 + 1)(2 + 1)$, or 12.

A further refinement of the intensities indicated in Figure 3-3c could be made by taking into account the relative numbers of protons in each group. The methyl triplet, for example, corresponds to three protons. The smaller peaks of this triplet should represent 3 divided by the sum of the ratios $(1 + 2 + 1)$ times the predicted intensity of 1, or $\frac{3}{4} \cdot 1$ or 0.75 proton. The center line of the triplet should have an intensity of $\frac{3}{4} \cdot 2$ or 1.5. The weakest lines of the middle methylene sextet should represent, on the same basis, $[2/(1 + 5 + 10 + 10 + 5 + 1)] \cdot 1$ or 0.063 proton. This refinement yields the approximate relative intensities for the peaks throughout the spectrum. These values are represented by the lines in Figure 3-2. It will be seen that, although the predicted and found intensities are in fair agreement, the distortions in intensities noted earlier are not predicted on this basis. Another refinement which will correct for these differences will be given in Chapter 4.

1-Nitropropane constitutes an $A_3M_2X_2$ system where $J_{AM} \approx J_{MX}$ and $J_{AX} \approx 0$.   Since there are a total of seven protons involved, this is called a seven-spin system. If the three absorption bands were closer together, this system would be called an $A_3B_2C_2$ type. If there were a reasonably large difference in energy among the three conformations about the carbon–carbon bonds, the system would become an $A_3'M_2'X_2'$ type. Slow rotation about the bond between the methylene groups could complicate this system even more, because the possible differences in chemical shifts of the protons on the two methylene groups might not be averaged out.

It is important to be aware that descriptions of spin systems are almost always only approximations. A system is usually described as the simplest approximation which is accurate enough to give the desired degree of agreement with the observed pattern.

First-order splitting diagrams are useful in helping to visualize spin patterns, but these diagrams should not be applied to systems which do not conform to the two simple conditions.

The spin systems emphasized so far have involved protons attached to the same or adjacent carbon atoms. Actually, protons on other adjacent atoms (such as C$\underline{H}$—O$\underline{H}$) can also couple. An example of this type of coupling is shown in Figure 3-4. The $-CH_2OH$ group constitutes an AX_2 system with $J_{AX} = 5$ cps. The hydroxyl proton thus appears as a 1:2:1 triplet centered at 196 cps (3.27 ppm), while the methylene protons appear as a 1:1 doublet centered at 251 cps (4.19 ppm). All of the spacings in the patterns due to these three protons are equal to J. The multiplets are symmetrically arranged about the chemical-shift positions.

Quite often, couplings of this type (—CHOH) are not seen because of inter- or intramolecular exchange. The splitting of the methylene group by the hydroxyl proton and the mutual splitting of the hydroxyl proton by the methylene protons occur because each group is affected by the different energy levels of the proton or protons in the other group. If these energy levels are changing rapidly, the influence is averaged out to zero. One process by which this averaging can take place is by exchanging the hydroxyl proton for another one having the other energy level. The rate at which the hydroxyl proton would have to exchange either inter- or intramolecularly in order to remove the observed splitting is approximately equal in exchanges per second to the value of J in cps [26]. This means that an exchange rate of 5 times a second would be sufficient to collapse the doublet and the triplet to single peaks. Quite clearly, under the conditions used to obtain the spectrum in Figure

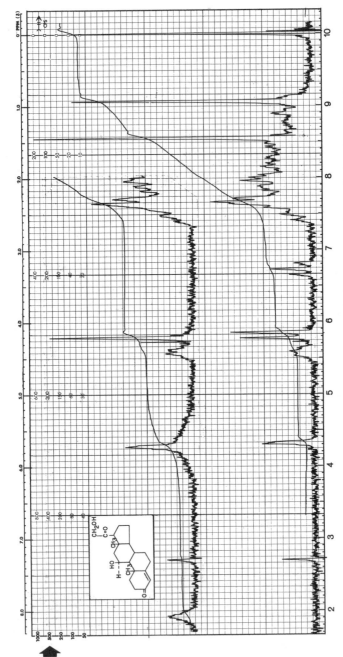

Fig. 3–4. Spectrum showing the coupling of a hydroxyl proton with the protons on the attached carbon atom. The upper curve was determined after the addition of a small amount of formic acid to the $CDCl_3$ solution.

3-4, the exchange was taking place at much less than 5 times per second. The addition of strong acid should increase this rate. The result of the addition of formic acid to the $CDCl_3$ solution is shown in Figure 3-4. Here both the multiplets have collapsed to give singlets, a broad band centered at about 340 cps for the hydroxyl, and a sharp peak at 253 cps for the methylene protons of the $-CH_2OH$ group. The broad band centered at 340 cps also represents the other hydroxyl proton in the molecule (at C-11) and the carboxylic acid proton in the added formic acid. The C-11 hydroxyl proton appears near 94 cps (1.56 ppm) in Figure 3-4 before the addition of the formic acid. The $H\overset{\text{O}}{\underset{\|}{C}}-$ proton of the formic acid appears near 485 cps (8.07 ppm). Although the exchange rate of the hydroxyl proton in the $-CH_2OH$ group is now faster than 5 times per second, the rate of exchange of all the protons in the solution is not yet fast enough to produce a single sharp peak.

Figure 3-5 shows the spectrum after D_2O exchange of the formic-acid-treated solution used in Figure 3-4. The signals in Figure 3-4 which were due to the hydroxyl groups and the formic acid are now essentially removed. A trace of formic acid which was not washed out can be detected at 485 cps. Even if rapid exchange were not occurring, the coupling in the $-CH_2OD$ group would be, as will be discussed later, only 0.154 of the corresponding $-CH_2OH$ coupling. This would be only 0.77 cps.

Dimethylsulfoxide has been recommended as a good solvent for use when it is desirable to see the coupling of a hydroxyl proton with a proton or protons attached to the carbon atom bearing the hydroxyl [70]. In the absence of strong acids or bases, the proton exchange rate is invariably quite slow in this solvent. Strong hydrogen bonding occurs with the solvent so that the hydroxyl proton peaks appear at fields lower (to the left) than 221 cps (3.68 ppm). Dimethylsulfoxide does not absorb below this position. The water present in the solvent shows up near 195 cps (3.25 ppm).

In Figure 3-5, the TMS signal is symmetrical, appears almost at 0 cps, and shows good ringing. The signal at 276 cps (4.60 ppm) is due to water left in the sample after the D_2O exchange, while the peak at 438 cps (7.30 ppm) is produced by chloroform. The angular methyl groups appear at 56 cps (0.93 ppm; C-18) and 87 cps (1.45 ppm; C-19). The presence of the C-11 β-hydroxyl (axial or essentially perpendicular to the plane of the carbon skeleton) shifts both the C-19 and C-18 methyl proton signals to the left by about

58

CHAPTER 3

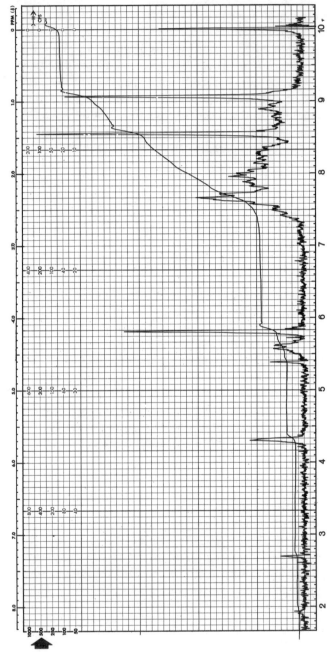

Fig. 3-5. Spectrum of the compound used in Fig. 3-4 after D₂O exchange.

15 cps. This large shift is characteristic of a 1,3-diaxial relation-ship of a hydroxyl to an angular methyl group in the steroids. Only one other hydroxyl group (C-8β) bears this relationship to both angular methyl groups, thus shifting both methyl proton signals to the left approximately (11 cps) this same extent [66]. The band at 264 cps (4.40 ppm) is due to the equatorial proton attached to the hydroxyl-bearing C-11 position. The pattern is approximately a quartet because the proton is coupled about equally to three other protons. This multiplet is more complex, however, than a first-order pattern. Similar complications will be discussed in Chapter 4. The relatively sharp C-4 olefinic proton signal appears at 341 cps (5.69 ppm). The protons on the saturated carbons which do not bear hydroxyl functions appear as broad background absorption be-tween 50 and 160 cps (0.83 and 2.67 ppm).

It should be remembered that the band position of a proton is very sensitive to hydrogen bonding, which causes a paramagnetic shift (to the left; downfield). This means that the hydroxyl peak would move to the right with increasing temperature. The tem-perature in the sample chamber of the Varian A-60 is about 32°C. Two different scans of the same compound may accidentally be run at two different temperatures if equilibrium is not established in both cases. This would mean that the signal due to the hydroxyl proton would appear at different positions in the two spectra.

All of the complications introduced into the spectrum by protons attached to oxygen are removed by the D_2O exchange technique. The coupling to other protons, the variations in position due to dif-ferences in hydrogen bonding, and the changes brought about by alteration of the exchange rate are all removed by the exchange.

A proton attached to a nitrogen or sulfur atom can, like those attached to oxygen, couple with a proton attached to an adjacent carbon atom in such groups as —CH—NH and —CH—SH. Typical coupling constants are given in Table 3-1. No coupling can be ob-served if rapid exchange (greater than J) is occurring. For this reason, this type of coupling, which is frequently found in amides, is seldom seen with amines.

The important facts presented so far which should be remem-bered about the signals due to protons attached to oxygen, nitrogen, and sulfur are summarized in Table 3-2.

Couplings which involve protons attached to nitrogen are further complicated by the possibility of coupling occurring with the nitro-gen nucleus itself. Spin–spin coupling can also occur with three other abundant, commonly occurring nuclei: H^2 (deuterium), F^{19}, and P^{31}. These magnetic nuclei are listed in Table 3-3 together with nuclei of less general interest.

The same multiplicity rules which have been discussed previously apply to all but four of the nuclei in Table 3-3. These exceptions are H^2, B^{11}, N^{14}, and O^{17}. The number of peaks expected when a group of equivalent protons is coupled with one of these four nuclei is given by the modifications of the $N + 1$ rule indicated in

TABLE 3-1

Proton–Proton Coupling Constants for Protons Attached to Oxygen, Nitrogen, and Sulfur

	J_{AB}(cps) (no exchange)
$CH_A OH_B$	4-7
$CH_A SH_B$	8
$CH_A NH_B$	not usually observed
$CH_A NH_B \overset{O}{\overset{\|}{C}} -$	5-9

TABLE 3-2

Effect of Conditions on Observed Coupling of Exchangeable Protons with Neighboring Protons

Y = oxygen, nitrogen, or sulfur

$-CH_A-YH_B$ (position of $-YH$ very sensitive to hydrogen bonding; dependent on solvent, temperature, and concentration).

Whether or not coupling is observed depends on J_{AB} and exchange rate; when rate in exchanges per second equals J_{AB} in cps, no coupling is observed.

Exchange with D_2O removes observable coupling; H_2O which is formed may appear near 282 cps.

TABLE 3-3

Nuclei of Interest in Organic Chemistry with Which Protons Can Couple

Nuclei which obey a modified multiplicity rule also cause broadening of the proton signals. The ranges of coupling constants listed for various arrangements are approximate. The data are taken from the Varian table [44], review by Lauterbur [10], and lecture by T. J. Flautt.

Nucleus (Y)	Natural abundance (%)	Modified multiplicity rule	J_{YH}, cps		
			Y—H	Y—C—H	Y—C—CH
H^1	99.98		280	0 to − 30	0 to + 15
H^2	0.02	$2N + 1$	40	0 to − 4	0 to + 2
B^{11}	81.17	$3N + 1$	30-190		
C^{13}	1.1		120-200	3-25	6
N^{14}	99.63	$2N + 1$	50	0	
O^{17}	0.04	$5N + 1$			
F^{19}	100		615	40-80	11-20
Si^{29}	4.70		120-250	6	6
P^{31}	100		180-700	5-15	5-15

Table 3-3. The ratios of intensities of the peaks expected when the signal due to a group of equivalent protons is split by these nuclei are also different from those indicated earlier (1:1, 1:2:1, 1:3:3:1, etc.). For a triplet produced by coupling to N^{14} or H^2, the peaks should have a ratio of intensities of 1:1:1. For a quintet produced by coupling to two such nuclei $(2N + 1)$, the intensities should be in the ratio of 1:2:3:2:1. A frequently observed example of this type is the quintet due to the group CHD_2— (see Figure 3-6b). Most deuterated solvents contain a few protons and, if a methyl group is present, there is usually some of the corresponding CHD_2— group in the solvent. The coupling constant between deuterium and a proton is about 0.154 of the coupling constant between corresponding protons. Protons attached to the same carbon atom under similar circumstances would be coupled to the extent of about −14.3 cps [−12.4 + (−1.9)]. The peaks in the quintet would thus be predicted to be separated by 0.154 · 14.3 or 2.1 cps. The value observed in Figure 3-6b for CHD_2COOD is about 2.3 cps.

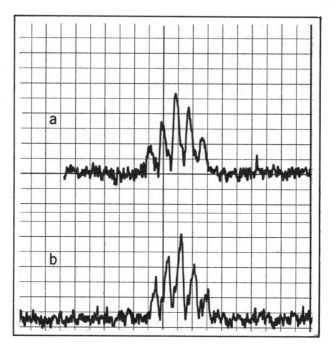

Fig. 3-6. The proton signal of CHD_2COOD in CD_3COOD. Curve b was determined by running the spectrum from left to right (increasing field); curve a was run from right to left (decreasing field).

The corresponding signal for the deuterium would be a doublet $(N+1)$ separated by 2.3 cps and located, if run on the usual scale, about 32 miles (about 51 Mcps) to the right in the spectrum. The observed distortion of intensities of the proton quintet from the predicted ratio $(1:2:3:2:1)$ is quite obviously not due to a small ratio of the difference in chemical shifts (about 51 Mcps) to the coupling constant (2.3 cps). The direction of the distortion is also not in agreement with this hypothesis. This is, in fact, a special effect (due to "saturation") which is dependent on the direction in which the spectrum is run [50]. The pattern in Figure 3-6a was obtained by decreasing, rather than increasing, the magnetic field. The peaks are now seen to be slanting towards higher field rather than lower field.

The approximate ranges of the coupling constants between protons and other nuclei are given in Table 3-3.

The coupling of a proton with the nitrogen nucleus is seldom seen either because of fast proton exchange or because of the non-spherical distribution of charge over the nitrogen nucleus. (A nonspherical charge distribution gives rise to an "electric quad-rupole moment.") The nonspherical charge distribution usually results in broadening of the peaks so that the signal of a proton attached to nitrogen may be spread over a very large range. The signals for amide protons, for example, are usually quite broad and are often difficult to detect. Their presence is most easily seen by inspection of the integration curve or by sighting along the absorption curve. Fast proton exchange (greater than 50 times per second because $J_{NH} \approx$ 50 cps) causes the three equally intense peaks to coalesce into a single sharp peak. The protons on most basic nitrogen atoms appear as very sharp signals because of rapid proton exchange. Other nuclei in Table 3-3 which have a non-spherical charge distribution are H^2, B^{11}, and O^{17}. The broadening of peaks which is due to nonspherical charge distribution de-creases rapidly with the separation of the nucleus and the protons with which the nucleus is coupled. It will be seen that all of those nuclei which cause broadening of peaks also fail to obey the $N+1$ rule.

The coupling of a nitrogen nucleus with a proton not attached directly to the nitrogen is quite weak. The coupling of protons with each other can, however, take place through C—N and N—H bonds (see Table 3-1).

The coupling of protons to Si^{29} and C^{13} can both be easily seen in the spectrum of tetramethylsilane (Figure 3-7). These spectra were obtained using liquid tetramethylsilane (no solvent). For convenience, the strong peak was placed in the center of the spec-trum. The rate at which the sample was spinning was different in the two scans. These spectra are actually the superimposed spec-tra of the following four isotopically different tetramethylsilane molecules: (a) $Si^{28}(C^{12}H_3)_4$ (88.3%), (b) $Si^{29}(C^{12}H_3)_4$ (4.5%), (c) Si^{30} $(C^{12}H_3)_4$ (3.0%), and (d) $C^{13}H_3Si^{28}(C^{12}H_3)_3$ (3.9%). Protons are the only magnetic nuclei present in (a) and (c). All of these protons are equi-valent and thus produce a single, sharp peak. This peak was allowed to go off scale in Figure 3-7 in order to show the minor peaks. About 5% of the molecules contain Si^{29}, with which the 12 protons can couple. The $N+1$ rule applies to the splitting caused by Si^{29} (see Table 3-3). The coupling constant is small, about 6 cps, so the signal due to the 12 protons is split into a doublet separated by 6 cps. The members of this doublet appear 3 cps on either side of the strong central peak. The doublet is relatively weak because of the low concentration of this type of molecule. About 4% of the molecules contain C^{13}, which also causes splitting according to the

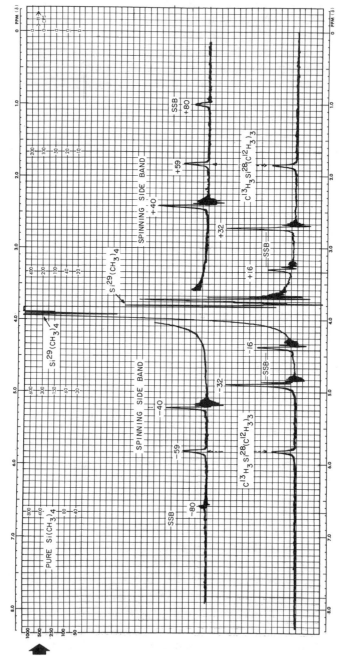

Fig. 3-7. Spectra of tetramethylsilane determined at two different sample spinning rates. The pure liquid was used without a solvent. The spectra were centered on the graph for convenience.

$N + 1$ rule. The coupling constant of C^{13} with the 12 protons is very large, 118 cps. The protons in the molecule containing the C^{13} thus appear as a doublet separated by 118 cps. The strong central peak appears in the middle of this doublet.

The doublets due to the coupling of the 12 protons to Si^{29} and C^{13} are called satellite signals. All signals due to protons attached to naturally occurring carbon have C^{13} satellite signals. Usually these peaks are lost in the noise. Very accurate integrations, however, require that these signals be taken into account if overlapping occurs with other peaks. When proton-containing substances are used as solvents, the C^{13} satellite signals will be fairly strong. Thus the regions within 60 cps on each side of the main solvent peaks will be of limited use. This can be seen in the solvent chart, Table 2-1.

The Si^{29} satellite appears in most spectra as a very small peak on the low-field side (left) of the TMS peak. The corresponding peak on the high-field side is obscured by ringing.

The signals labeled "spinning side band" (SSB) are a result of an interaction of the spinning sample tube with a slightly inhomogeneous magnetic field. These signals occur at multiples of the rate at which the sample is spinning. The usual spinning rate is about 30 rps. The identification of the spinning side bands is easily established by comparison of the two scans made at different sample spinning rates. The intensity of these spinning side bands can be used to judge the operation of the instrument. The smaller these peaks are, the better the instrument is operating. These peaks, like the C^{13} satellite signals, are usually lost in the background noise.

An interesting complication arises with many primary amides. The C—N bond possesses enough double-bond character to hold one proton in a cis configuration and the other in a trans configuration. Thus the two protons in

$$\underset{/}{\overset{O_{\diagdown}}{}}C = N\underset{\diagdown H}{\overset{/H}{}}$$

bear a different spatial relationship to the carbonyl function and consequently their signals will appear as two broad peaks at different magnetic field strengths. An example of this was shown in Figure 2-12.

The factors which influence the spin patterns of signals due to protons attached to nitrogen are summarized in Table 3-4. In Table 3-5, the characteristic appearances of these signals are related to the various functional groups.

The nonequivalence of the primary amide protons brings up a very important general consideration. That is, that the magnetic equivalence or nonequivalence of protons is not always consistent with the organic chemist's concept of chemical equivalence or nonequivalence. One class of such protons are, like the primary amide protons, different because of different environments which result from restricted rotation about partial double bonds. Common examples of this type of group are given in Table 3-6. Each of these groups can behave like a mixture of cis and trans isomers. If the two isomers are not rapidly interconverted, both may show up in the spectrum as distinct compounds. The interpretation of spectra of such compounds can be quite baffling if this possibility is overlooked. An example is given in Figure 3-8, which shows the spec-

TABLE 3-4

Influence of Various Factors on Appearance of Signal Due to Proton Attached to Nitrogen

Note that primary amides can give two broad peaks (see Table 3-5).

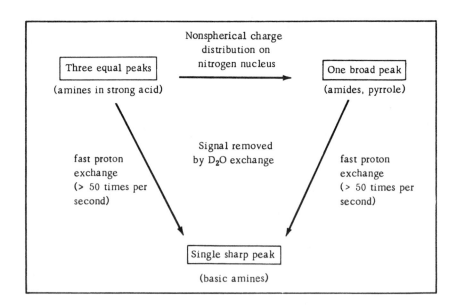

TABLE 3-5

Characteristic Appearance of Signals Due to Protons Attached to Nitrogen in Various Functional Groups

Basic Amines

$-CH_A-NH_B$ Usually produce single, sharp peak due to rapid exchange.

$\overset{\oplus}{R_3}N\overset{\ominus}{H}X$ In strong acid, amines may give three rather broad, equally intense peaks due to coupling with the nitrogen nucleus.

Pyrrole

Proton on the nitrogen appears as extremely broad band.

Amides

$\overset{O}{\overset{\|}{R C N H_2}}$ Often give two broad bands because of restricted rotation about the C—N bond. The broadness is due, in part, to coupling with nitrogen.

$\overset{O}{\overset{\|}{R C N H_A C H_B}}$ J_{AB} is usually in the range of 5 to 9 cps. This coupling is often observed, which means that the exchange rate is much less than 5 to 9 times per second. Splitting may be confused with nonequivalence due to restricted rotation.

---

trum of N-nitrosopiperazine. Piperazine itself $\left(HN \overbrace{} NH \right)$ shows only two peaks, one for the rapidly exchanging protons attached to nitrogen and one for the eight equivalent ring protons. The nitroso group exerts both a spatial effect and an electron-withdrawing effect. Both of these effects cause a downfield shift (to the left) of the signals due to the closest methylene protons. The partial double-bond character of the N—N bond stabilizes the two equivalent forms:

HN ⬡ N—N=O   and   HN ⬡ N—N=O

TABLE 3-6

Some Common Systems Which May Show Restricted Rotation About Single Bonds Having Partial Double-Bond Character

The signal due to the protons on the same side as the oxygen are displaced further downfield (to the left) than the signal due to the protons on the other side. The overall effect is that two different $A_2'X_2'$ patterns are seen. If rapid rotation occurred about the N—N bond, the difference would be averaged out, and only one $A_2'X_2'$ pattern would be seen. This could be brought about by increasing the temperature.

While restricted rotation about a partial double bond can give rise to two possible isomers, restricted rotation about a single bond in an aliphatic system can give rise to three isomers or conformers. This situation is illustrated in Figure 3-9. The steric interaction of the substituents on the two carbon atoms tend to stabilize the three forms, or conformations, in which the substituents are farthest apart. Both the rate of rotation about the carbon—carbon bond and the relative populations in the three conformations depend on the bulkiness and nature of the substituents and upon the temperature. An increase in temperature not only increases the rate of rotation about the carbon—carbon bond, but also causes the conformations to be more evenly populated. The two protons in Figure 3-9 may be different from each other in each of the three conforma-

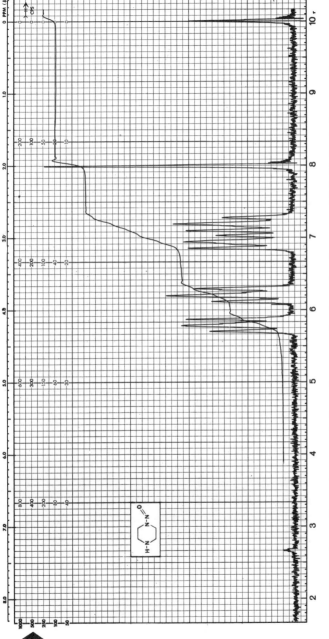

Fig. 3-8. Spectrum illustrating nonequivalence due to restricted rotation about a partial double bond. Determined in CDCl₃.

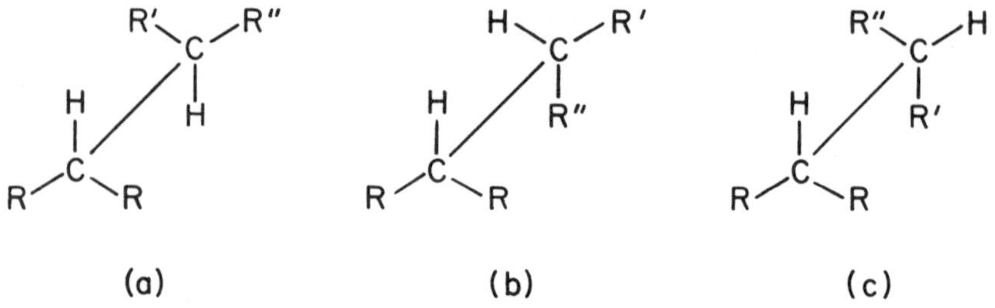

Fig. 3-9. Rotational isomers of a compound of the type CHRRCHR'R". If there were slow rotation about the carbon–carbon bond, this could constitute three AB (or AX) types. With fast rotation and equally populated conformations, this would be one AB (or AX) type.

tions. With fast rotation about the carbon–carbon bond, the system would be simply an AB or AX type because the environments of each of the protons would be averaged over the three isomers. The angularly dependent coupling constant between the two protons would also be averaged. Both the average chemical shift and the average coupling constant would be weighted according to the populations of the three isomers. If the rotation were very slow, however, each of these conformations could appear as separate components in the NMR spectrum, giving rise to three AB or AX systems instead of one. There could be three coupling constants. The coupling constant for (a), with the dihedral angle of 180°, would be greater than that for (b) or (c).

The discussion of the example in Figure 3-9 emphasizes the point that there are two separate aspects of restricted rotation which must be considered. The first is the rate of rotation about the bond, and the second is the population of the various forms. If the energy barriers to rotation are large enough to prevent fast rotation, the possibility of unequal populations of the forms would be easily suspected. However, even though there is fast rotation, the populations of the various forms may not be equal. In simple molecules, NMR can often serve very elegantly in studies of hindered internal rotation [51].

The same time dependency discussed in conjunction with the rate of exchange of protons attached to oxygen, nitrogen, and sulfur applies to the rate of rotations about bonds. If the environments of two or more protons are changed at a rate in cycles per second equal to or faster than the differences in chemical shifts in cycles per second, the signals merge into a single peak.

Problems in spectral interpretations resulting from oversight of restricted rotation about a single aliphatic bond having no double-bond character are not very frequently encountered. The possibility should always be kept in mind, however.

A very common cause of nonequivalence of methylene protons is restricted rotation due to rigid ring systems. At room temperature, the two stable chair conformations of cyclohexane are being interconverted sufficiently rapidly to average out the difference in chemical shift expected for axial and equatorial protons. Thus only a single, sharp peak is observed at 86 cps. When the cyclohexane system is made rigid by suitable fusion to other rings, the difference between axial and equatorial protons is no longer averaged out. In steroids, for example, the ring methylene protons usually give a broad background absorption band between 50 and 150 cps. Much complexity also arises in this band because of complex spin-spin interactions. Very few of these patterns are first-order. Magnetic nonequivalence of methylene protons in rigid ring systems causes little trouble, since the organic chemist has learned to expect chemical nonequivalence as well.

One very commonly occurring case of nonequivalence of methylene protons arises when a nearby carbon atom has three different groups attached to it. For example, in the system $-CH_2CR^1R^2R^3$, the two methylene protons may be nonequivalent. This is only a permissive condition for nonequivalence. It often happens that such protons are equivalent. The nearby carbon atom may, or may not, be an optically active asymmetric center [52, 54] since only three, rather than four, different substituents are required. Thus in

$$\underset{\underset{R^2}{|}}{\overset{\overset{R^1}{|}}{BrCH_2C}}-CH_2Br$$

the methylene protons in each $BrCH_2$ group may be nonequivalent because there are three different substituents on the adjacent carbon. This could then result in two superimposed AB patterns. This molecule would not be optically active, however, because there are only three, not four, different substituents on the carbon atom. The effect of a carbon atom with three different groups attached can extend further than the next atom. In Figure 3-10 the nonequivalence of the methylene protons in the $-CH_2O\overset{\overset{O}{||}}{C}CH_3$ group results in a typical AB pattern in the 250-cps region. In this example, the nonequivalence is due to the three different substituents on the adjacent carbon atom, which is an asymmetric center. Each proton signal is split into a doublet by the other proton, giving rise to the pair of doublets at 236 and 247 cps and at 258 and 269 cps. The coupling between the protons in the $-CH_2OAc$ group is thus 11 cps, in agreement with the typical values given in Table 2-7a.

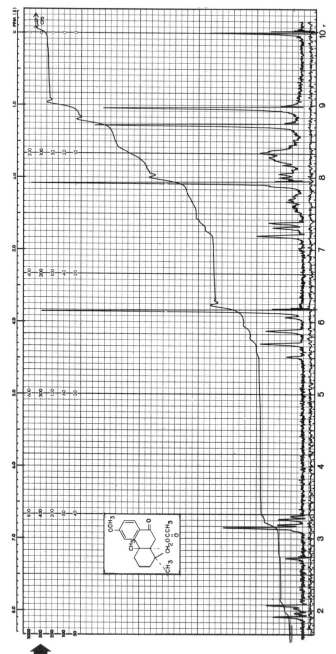

Fig. 3-10. Spectrum illustrating nonequivalence due to three substituents on the adjacent carbon atom.

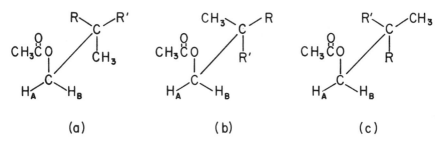

(a)   (b)   (c)

Fig. 3-11. Stable conformations of a system which shows "three-substituent" nonequivalence. It is possible that the difference in the environments of the two methylene protons may not be averaged out even with fast rotation and equal populations in the three stable conformations.

The TMS signal in Figure 3-10 is normal, and the chloroform signal can be seen. The protons in the two tertiary methyl groups, the methyl of the acetyl, and the methoxy group can be easily identified by their position, sharpness, and intensity. These are A_3 systems. The aromatic protons constitute an ABX pattern which appears in the 400 to 500 cps region. The three sharp peaks near 160 cps (2.67 ppm) are part of an ABC (or ABX) pattern produced by the two protons adjacent to the ketone and the angular proton at C-5. These more complex patterns will be discussed in Chapter 4. The remaining six protons on the saturated carbon atoms give the background absorption between 85 and 150 cps (1.42 and 2.50 ppm).

Nonequivalence of methylene protons due to three different substituents on a nearby carbon atom does not necessarily result from restricted rotation or unequal populations of the conformations [13]. Even with fast rotation and equal populations, the differences in the environments of the two protons may not be averaged out. This can be seen by consideration of the three stable conformations in Figure 3-11. When H_A is placed between R— and CH_3— (Figure 3-11a), it is influenced by the effect of $CH_3\overset{O}{\overset{\|}{C}}O$— being between R— and R'—. When the group is rotated so that H_B is now in the equivalent position (between R— and CH_3—; Figure 3-11c), H_B is influenced by $CH_3\overset{O}{\overset{\|}{C}}O$— being between R'— and CH_3—, not between R— and R'—. Thus the presence of three substituents on a nearby carbon atom may introduce asymmetry which cannot be averaged out by rapid rotation. In some cases the differences in chemical shifts produced by this asymmetry are too small to be detectable.

TABLE 3-7

Conditions Which May Produce Unexpected Nonequivalence of Protons

I.	Restricted (or prohibited) internal rotation about a
A.	double bond
B.	partial double bond
C.	single bond:
	1. because of bulky substituents
	2. because of rigid ring systems
II.	Restricted inversion of a nitrogen atom
III.	Asymmetry when a nearby carbon atom carries three different substituents (possible that the carbon atom may not be an optically active asymmetric center)

TABLE 3-8

Some Features of Concern in the Interpretation of Spin–Spin
Patterns

I. Many spin system designations are only approximations. The best description is the simplest one which gives the desired agreement with the observed pattern.

II. First-order patterns and the relative intensities of the peaks can be pictured by the use of first-order splitting diagrams.

III. Methylene protons which are chemically equivalent may be magnetically nonequivalent because of restricted rotation, restricted inversion about a nitrogen atom, or because of molecular asymmetry.

IV. Differences in chemical shifts among protons is averaged out if the environments of the protons are interchanged at a rate equal, in times per second, to the difference in chemical shift in cycles per second.

V. Coupling can, and does, occur with other magnetic nuclei. Of particular concern are H^2, N^{14}, F^{19}, and P^{31}.

The point to remember is that methylene protons near a carbon bearing three different substituents may be nonequivalent.

"Three-substituent" nonequivalence of methylene protons can also extend to groups other than protons attached to the methylene carbon. For example, in a group of the type

$$-\underset{\underset{CH_3}{|}}{\overset{\overset{CH_3}{|}}{C}}-\underset{\underset{R^3}{|}}{\overset{\overset{R^1}{|}}{C}}-R^2,$$ the

two methyl groups may be nonequivalent. They would then show up as two separate sharp signals. In a group such as

$$-\underset{\underset{CH_2}{|}}{\overset{\overset{CH_2}{|}}{\underset{|}{\overset{|}{C}}}}\underset{CH_3}{}-\underset{\underset{R^3}{|}}{\overset{\overset{R^1}{|}}{C}}-R^2,$$

the two methylene groups may be different from each other. The terminal methyl groups, being further removed, would probably, although not necessarily, be equivalent. Each of the methylene groups would be spin–spin coupled to the methyl groups. Since the difference in chemical shift between the methylene protons and the methyl protons would probably be small compared to $J_{CH_3 - CH_2}$ (6-8 cps), the methylene proton absorption could indeed be complex.

Thus far, no cases of the nonequivalence of the three protons in a methyl group ($-CH_3$) have been reported. Their nonequivalence cannot be brought about by three different substituents on a nearby carbon atom. Nonequivalence could be introduced, however, by slow rotation about the bond attaching the methyl group to the molecule.

Table 3-7 summarizes the conditions which may produce unexpected nonequivalence of protons.

Three of the causes of nonequivalence listed in Table 3-7 may be averaged out by increasing the temperature, thus increasing the rate of rotation or inversion. These causes are: IB (restricted rotation about a partial double bond), IC (restricted rotation because of bulky substituents), and II (restricted inversion at a nitrogen atom). The other causes of nonequivalence cannot be averaged out by a change in temperature.

The essential features of concern in the interpretation of first-order spin–spin patterns which have been discussed in this chapter are outlined in Table 3-8.

The distortions of the spin patterns which result when a system fails to conform to one or both of the first-order conditions will be considered in the next chapter.

Chapter 4

Higher-Order Spin Patterns and Multiple Resonance*

The next concern is the appearance of spin patterns which are not first-order. Deviations from first-order patterns can be made either by having $\Delta\nu < 6J$ or by having unequal coupling constants between the groups of protons. The resulting multiplets will be referred to as "higher-order" patterns. The general effect of departures from these two conditions will be discussed first.

Three changes in the spin patterns may be observed when the two first-order conditions are not met: (1) the simple ratios of peak intensities are distorted, (2) extra peaks appear, and (3) the spacings of the peaks may become unequal.

These effects can be seen in the series of patterns in Figure 4-1. The changes result from decreasing the chemical shift between the two groups of methylene protons while holding the coupling constant at about 7 cps. The patterns range from A_2X_2 in (a) to A_4 in (e).

Even in (a) it can be seen that the inner peaks of the triplets are somewhat stronger than the outer peaks. This increase in intensity of the inner peaks at the expense of the outer peaks continues as the two bands are brought closer together. In (b) the inner peaks are almost as intense as the middle peaks. In general, the relative intensities of peaks in spin multiplets tend to slant upwards in the direction of the other band. One practical application of this distortion in intensities is that simple inspection of one band can be used to determine if the other band is at higher or lower field. Even when the ratio of $\Delta\nu$ to J is 20:1, the slanting of the intensities can be easily detected.

In (b) the outer, weaker peaks of the triplets have broken up into doublets. If the changes which are expected on departure from the first-order patterns are kept in mind, (b) is still easily identified as resulting from approximately an A_2B_2 type. (See the dis-

---

*See references 4, 6, 45, 46 for higher-order spin patterns; reference 12 for multiple resonance.

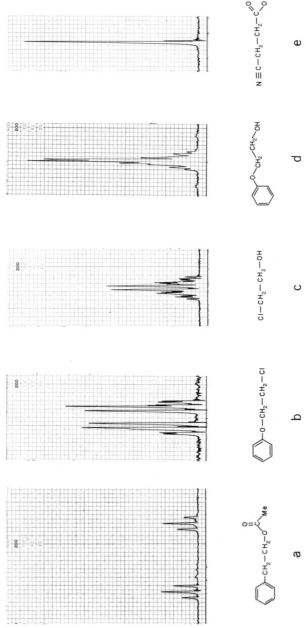

a

$CH_2-CH_2-O-\overset{\overset{\displaystyle O}{\|}}{C}-Me$

b

$\langle \rangle-O-CH_2-CH_2-Cl$

c

$Cl-CH_2-CH_2-OH$

d

$\langle \rangle-O-CH_2-CH_2-OH$

e

$N\equiv C-CH_2-CH_2-\overset{\overset{\displaystyle O}{\|}}{C}-O-Me$

Fig. 4-1. Effect of decreasing $\Delta\nu_{AB}/J_{AB}$ on the spin multiplet due to an A_2X_2 system. These patterns range from A_2X_2 in (a) to A_4 in (e). Although patterns b, c, and d are approximately the A_2B_2 type, detailed study would be required to determine whether or not these are actually A_2B_2 or $A_2'B_2'$ patterns. See the discussion on p. 80. The chemical shifts can be determined by inspection only in the first-order patterns, (a) and (e). The coupling constant can also be measured in (a). However, neither $\Delta\nu_{AB}$ nor J_{AB} can be determined by inspection of the higher-order patterns, b, c, and d. All of these patterns are reprinted from the Varian Catalogs [7,8] by permission of the copyright holder, Varian Associates.

cussion below.)   In (c) the middle line of the triplet has also broken up, and the outer line has broken up still further.

The peaks in (b) and (c) are no longer separated by J. In fact, care must be taken not to incorrectly assign J in these higher-order systems.

As $\Delta\nu$ decreases, the distortion of intensities continues and eventually the very weak outer lines are lost. The central strong lines merge. The final result is a single sharp peak for four equivalent protons ($\Delta\nu \approx 0$) as in Figure 4-1e.

The successive changes brought about in spin patterns by decreasing $\Delta\nu/J$ are summarized in Table 4-1.

Deviations due to unequal coupling constants are not as great a problem as they would first seem. If the spin systems outlined in Table 4-2 are considered, no confusion arises in any of the systems consisting of fewer than four protons. The problem is not involved in the two-spin systems because there is only one coupling constant. The three coupling constants involved in the ABC, ABX, and AMX systems are not likely to be assumed to be equal. In the A_2B and A_2X systems, the patterns are characteristic of $\Delta\nu_{AB \text{ (or AX)}}/J_{average}$ (see Figure 4-4), and no differences are observed when the coupling constants are not equal. The smallest spin systems which are likely to cause difficulty because of unequal coupling constants are the $A_2'B_2'$ and $A_2'X_2'$ types. These systems are frequently encountered, and the inequality of the coupling constants must be considered.

TABLE 4-1

Sequence of Changes in Spin Pattern Brought About by Decreasing the Ratio of the Difference in Chemical Shifts ($\Delta\nu$) to the Coupling Constant

1. Relative intensities of peaks toward other part of pattern increases at expense of peaks away from the other part of the pattern.

2. Some single peaks break up into multiplets.

3. Weaker outer peaks are lost in the noise.

4. Center peaks merge.

5. Center peak sharpens.

6. All outer peaks are lost in the noise.

TABLE 4-2

Commonly Occurring Spin Systems

The primed systems are those in which each proton in one group is not coupled equally to each and every proton in the second group. The higher-order systems are in bold type. All of the other systems in this table are first-order.

A					
A₂	**AB**	AX			
A₃	A₂B	A₂X	**ABC**	**ABX**	AMX
	A₂′B′	**A₂′X′**			
A₄	**A₃B**	A₃X	**A₂B₂**	A₂X₂	
	A₃′B′	**A₃′X′**	**A₂′B₂′**	**A₂′X₂′**	
A₅	**A₂B₃**	A₂X₃	**ABC₃**	**ABX₃**	
	A₂′B₃′	**A₂′X₃′**			

Many nonsymmetrical 1,2-disubstituted ethanes (XCH_2CH_2Y) are actually common examples of $A_2'B_2'$ and $A_2'X_2'$ systems [77]. The inequality of the coupling constants in this group arises whenever the three stable conformations are not equally populated. (See the discussion on p. 68.) When one conformation is more highly populated, the dihedral angles between one of the methylene protons and each of the two protons on the adjacent carbon atom are, on the average, different from each other. This means, then, that two different coupling constants may be involved. A detailed analysis of the patterns in Figure 4-1 would have to take this possibility into account. In general, the triplet nature of the two halves of the aliphatic XCH_2CH_2Y pattern is not obscured by the inequality of the coupling constants between the protons in the two groups. The chief variation which is usually introduced is in the fine structure of the middle peak [77]. It should be noted that the involvement of two different coupling constants in these cases is not usually due to slow rotation about the carbon–carbon bond but rather to the differences in the populations in the three stable conformations. Slow rotation would complicate the situation even more.

An illustration of the possible dramatic difference between two different $A_2'B_2'$ systems (or between approximately an A_2B_2 and an $A_2'B_2'$ system) is shown in Figure 4-2.

With this background of the effects of departures from the two first-order conditions, attention will now be focused on the individual spin systems listed in Table 4-2. It should first be noted that this table encompasses an enormous range of complexity. While equivalent protons can only produce a single peak, the $A_2'B_3'$ system can give 210 possible peaks [53]. Several of these systems, including the $A_2'B_3'$ type, are too complex to discuss in general fashion. Methods of dealing with these more complex systems will be discussed later. The objective at this point will be to make as many practical observations as possible about the simpler types of spin systems. The comments which will be made about the various systems are summarized in Table 4-3.

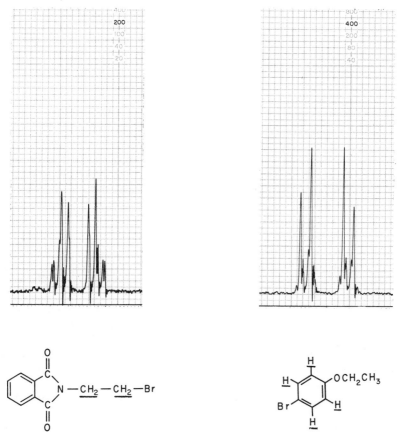

Fig. 4-2. Comparison of an $A_2 B_2$ (left) and $A_2 'B_2 '$ (right) system. Reproduced by permission of the copyright holder, Varian Associates.

TABLE 4-3

Prediction of Spin Patterns for Common Systems [6,45,46]
The determination of the spin system from the pattern is given in Table 6-1

System*	Maximum number of lines†	Comments
A_N	1	Equivalent nuclei give a single peak even though they are strongly coupled.
$A_N X_P$	For A: $P+1$; for X: $N+1$	Each band is symmetrical and has peaks separated by J_{AX}; peaks have relative binomial intensities. ‡
AMX	12	Each band has two or four equally intense peaks. Also see ABX below.
AB	4	$\dfrac{\text{Intensities of weaker outer peaks}}{\text{Intensities of inner peaks}} = \dfrac{\text{Separation of inner peaks}}{\text{Separation of outer peaks}}$; $\Delta\nu^2_{AB} = (\text{separation of peaks 1 and 3})^2 - J^2_{AB}$
A_2B A'_2B'	8 + 1	B: slanting** quartet. A: pair of doublets; outer doublet has greater separation. For A'_2B', pattern is a function of $\Delta\nu_{AB}/J_{AB_{average}}$
ABC	15	No good generalizations.
ABX	12 + 2	X: two (or three) doublets symmetrical about center of X band. Separation of strongest doublet is equal to $J_{AX} + J_{BX}$ Becomes a triplet if: $J_{AX} = J_{BX}$ or $J_{AB} \gg \Delta\nu_{AB} + \frac{1}{2}(J_{AX} - J_{BX})$. AB: two typical AB patterns (may be superimposed) J_{AB} occurs 4 times.
A_2B_2 $A'_2B'_2$	14 24	Band symmetrical about center but not about center of each half.
$A'_2X'_2$	20	Band symmetrical about center and also about center of each half. Each half: strong doublet plus two symmetrical quartets each having more intense inner lines.
A_3B	14 + 2	A: two bands, outer one weaker. Separation of these two bands is less than J_{AB}

*The primed systems are those in which each A is not coupled equally to each B or X.
†The second number given indicates the number of lines which in nearly all cases are very weak or absent.
‡"Relative binomial intensities": 1:1, 1:2:1, 1:3:3:1, 1:4:6:4:1, etc.
**"Slanting": intensities increase in direction of other portion of the pattern.

The spin systems in Table 4-2 which involve only equivalent nuclei all give rise to only one peak. Those systems which involve only A and X and which are equally coupled (unprimed) obey the simple multiplicity rules. The first system to be discussed will be the AB type.

Every aspect of the AB system can be easily treated in exact terms. Reference to Figure 4-3 will show how J, $\Delta\nu_{AB}$, and the ratio of integrated areas under the absorption peaks can be derived. Regardless of the ratio of $\Delta\nu$ to J, the doublets are separated by exactly J. This is not true in the more complex spin systems. If the peaks are numbered in the order in which they appear, the relative intensities of peaks 1 and 2 and peaks 4 and 3 are in the same ratio as the separation of the inner peaks (2–3) is to the separation of the outer peaks (1–4). If the measured value of J is placed on one side of a right triangle, and the distance between peaks 1 and 3 on the hypotenuse, the other leg of the triangle represents $\Delta\nu_{AB}$. Once $\Delta\nu_{AB}$ is found, measurement of $\frac{1}{2}\Delta\nu_{AB}$ from the center of the pattern will give the chemical shifts of the two nuclei. These relationships should be mastered completely.

Often only half of the AB pattern can be easily identified, the

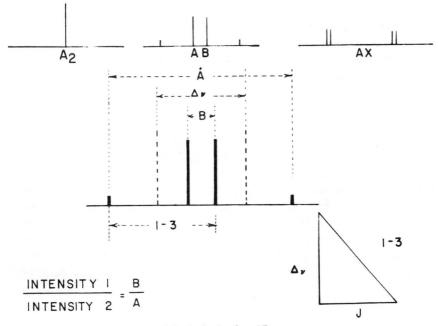

Fig. 4-3. Analysis of an AB system.

other half being in a region of overlapping bands. From the relationships above, it can be shown that the distance (1–4) from the
outer, weaker peak to the outer peak of the other doublet is given by

$$(1-4) = 2J/(1-\text{Area under outer peak}/\text{Area under inner peak})$$

If $\Delta\nu_{AB} = \sqrt{3}\,J$, the peaks for an AB pattern will all be equally
separated and have intensities of $1:3:3:1$. This pattern might
at first be thought to be indistinguishable from the A_2 band of an
A_2X_3 pattern. It will be noted, however, that the A_2 band of an A_2X_3
pattern would normally have at least a small distortion in intensities such that the other peak of the quartet would be smaller than
the inner peak. For the AB pattern, this distortion shows up in the
ratio of the doublets, so that the outer peaks of the "quartet" would
be equally intense. Further inspection should also reveal the possible presence of the X_3 triplet separated by J and showing a similar
distortion of intensities. A change of solvent might also be useful
here since $\Delta\nu$ is often relatively sensitive to solvent effect while J
is relatively insensitive.

The expected distortion of intensities in multiplets for patterns
approximating first-order spectra can be estimated using the relationships indicated for the AB case. The first-order diagram is
drawn as indicated earlier. A line is used to indicate each group
of protons. This line is then split once for each proton with which
that group of protons is coupled. The ratio of intensities of the
resulting doublet is determined using the AB rule. On completion
of this diagram, with care taken to adjust the ratio of intensities
each time a doublet is formed, fair agreement can be obtained with
the observed intensities.

All of the one-spin and two-spin systems can thus be easily and
completely analyzed. Among the three-spin systems, the A_3, $A_2'X'$,
A_2X, and AMX types give first-order spectra. The inequality of the
two coupling constants in the $A_2'X'$ system causes no change in the
pattern. The intensities of the peaks in AMX patterns can be calculated using the AB intensity rule. The remaining three-spin systems are the A_2B, $A_2'B'$, ABC, and ABX types.

The pattern due to an $A_2'B'$ system is a function only of the ratio
$\Delta\nu_{AB}/J_{AB\,\text{average}}$. The pattern is unaffected by an inequality of the two
coupling constants. The changes which take place as this ratio is
decreased are illustrated in Figure 4–4. In going from the A_2X system to the A_3 system, the expected distortion of intensities, appearance of extra peaks, and occurrence of unequal spacings can be
observed. The extra peaks arise from an apparent splitting of each

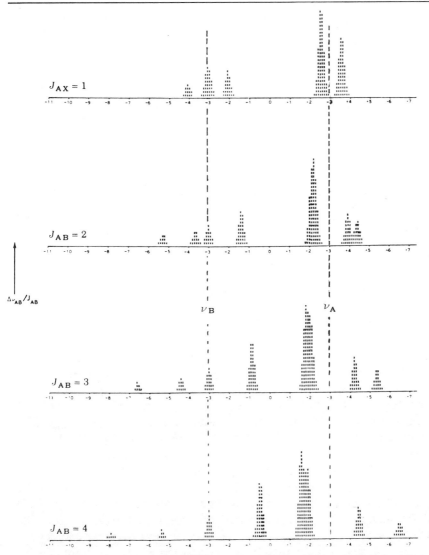

Fig. 4-4. Calculated patterns which show the effect of holding $\Delta\nu_{AB}$ constant (6 units) and increasing J_{AB} in an $A_2 B$ or $A_2' B'$ system. Only the top pattern ($A_2 X$) is first-order. These patterns are reprinted from the text of Wiberg and Nist [65] by permission of the copyright holder, W. A. Benjamin, Inc.

of the two peaks in the A_2 pattern, a splitting of the middle peak of the B triplet, and the appearance of an entirely new weak peak on the outside of the A_2 pattern.

A mental process which often helps in identifying higher-order spectra, providing the patterns are not overlapping, is as follows: (1) the closest spaced peaks in each portion are brought together, and (2) the distortion of intensities is corrected. This process would, for example, help to easily recognize the patterns in Figure 4-4 as being due to an A_2B system.

The extra peaks which are found in higher-order spectra are not intuitively explainable. The processes which occur to cause the absorption of energy are, of course, controlled by quantum mechanical laws. It is fortunate that the situation is so simple when the two first-order conditions are met.

The ABX type is the simplest system which permits a full discussion of the important basic concepts and the potential pitfalls involved in higher-order systems. Before these topics are considered, both the general nature of the ABX system and the extraction of the coupling constants and chemical shifts from the ABX pattern will be discussed [47-49,55].

The ABX system involves three coupling constants (J_{AB}, J_{AX}, and J_{BX}), which can be either positive or negative, and three chemical shifts for the three different nuclei. These six variables introduce a fair amount of complexity into the pattern. As a result, the discussion must involve a large number of details. The calculated ABX pattern shown in Figure 4-5 will be used as an illustration.

The ABX system is higher-order because the difference in chemical shift between the A and B nuclei is less than six times the coupling constant between A and B. If $\Delta\nu_{AB}$ were at least six times J_{AB}, this would be designated as an AMX system. The AMX system would obey the first-order rules. Each of the signals could be split into a doublet by coupling to one of the other nuclei, and each of these peaks could be further split into doublets by coupling to the other proton. Thus each of the protons could be represented by four peaks, giving a maximum of 12 possible peaks. The chemical shifts and coupling constants could be measured directly from the AMX pattern. As $\Delta\nu_{AM}$ is decreased to less than six times J_{AM}, two extra peaks may appear in the X pattern, giving a total of 14 possible peaks. The chemical shift of the X nucleus (ν_X) will still be at the center of the X pattern, but the chemical shifts of A and B cannot be determined by inspection. As will be seen later, only two other values can be measured directly from the ABX pattern, namely J_{AB} and $J_{AX} + J_{BX}$. Actually, the signs of these quantities cannot be determined. Quantities whose magnitudes are known but whose signs are not known are commonly placed in straight brackets as

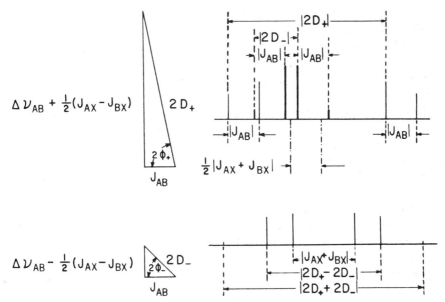

Fig. 4-5. Graphical representation of an ABX system in which $\Delta\nu_{AB} = -6$, $J_{AB} = 2$, $J_{AX} = -6$, and $J_{BX} = 2$. The AB pattern is shown in the upper section and the X pattern in the lower section. The notation follows that used by Pople, Schneider, and Bernstein [47]. Data from Wiberg and Nist [64].

$|J_{AB}|$. This means that J_{AB} may be either plus or minus the quantity in the brackets. This notation is employed in Figure 4-5 and will be used throughout the following discussion.

In Figure 4-5 the expected peaks are represented by vertical lines. The X proton is represented by the six vertical lines shown in the lower section of Figure 4-5. The A and B protons are represented by the eight peaks (lines) shown in the upper section. Examination of the AB portion (upper section) reveals two overlapping, typical AB patterns. (Figure 4-3 shows a single AB pattern.) One of the AB patterns is represented by the lighter lines and the other AB pattern is represented by the heavier lines. The outer pairs of peaks in each of the two typical AB patterns are separated by $|J_{AB}|$. The entire ABX pattern can consist of a maximum of 14 peaks, six for the X proton and eight for the A and B protons.

For convenience, Pople, Schneider, and Bernstein in their discussion of the mathematics involved in the ABX system [47] designated the distance between the first and third peaks of the two different AB patterns as $2D_+$ and $2D_-$ (see Figure 4-5, upper section). These same symbols will be retained in this book, so that the reader may easily follow either the mathematical treatment given by Pople, Schneider, and Bernstein or the graphical treatment given here.

The changes in the ABX pattern which are brought about by variations in the chemical shifts and coupling constants can be readily understood on the basis of fairly simple graphical relationships. While this discussion will be helpful, it is not essential that the reader consider all of the details. He may choose at this point to master only the information summarized in Table 4-5 and the comments about the ABX pattern in Table 4-3 and then go directly to the ABC system on page 96. The detailed analysis of an ABX pattern can be made following the procedure given in Table 4-4.

The quantities of interest which can be determined from the AB portion of the ABX pattern are $|2D_+|$, $|2D_-|$, $|J_{AB}|$, and $\frac{1}{2}|J_{AX}+J_{BX}|$. The first quantities, $|2D_+|$ and $|2D_-|$, are the separations of the first and third peaks of the two typical AB patterns. The value of $|J_{AB}|$ is equal to the separation of the outer peaks of each of these two AB patterns. The centers of the two AB patterns are separated by $\frac{1}{2}|J_{AX}+J_{BX}|$.

The values of $|2D_+|$ and $|2D_-|$ are also involved in the X pattern (lower section of Figure 4-5). Two of the X peaks are separated by $|2D_+ - 2D_-|$ and two other peaks are separated by $|2D_+ + 2D_-|$. The only other separation of interest in the X pattern is the distance $|J_{AX} + J_{BX}|$ between the two strong peaks. These strong peaks are separated by exactly twice the separation of the centers of the two AB patterns in the AB portion.

The relationships among these quantities can be represented by two triangles (left-hand side of Figure 4-5). These triangles are very similar to the triangle used earlier for the AB case (see Figure 4-3). The base of each triangle is J_{AB}. The hypotenuse of each triangle is again the separation ($2D_+$ or $2D_-$) of the first and third peaks of an AB pattern. The vertical sides of the two triangles again involve $\Delta\nu_{AB}$ but, because of the extra variables, this value is modified by $\frac{1}{2}(J_{AX} - J_{BX})$. This quantity is added to $\Delta\nu_{AB}$ in one triangle and subtracted from $\Delta\nu_{AB}$ in the other triangle.

The general procedure for the analysis of the ABX pattern will be indicated now. The details of this analysis are given in Table 4-4. The extraction of the coupling constants and chemical shifts starts with the direct measurement of $|J_{AB}|$, $|J_{AX} + J_{BX}|$, $|2D_+|$, and $|2D_-|$. Only the magnitudes of these quantities can be determined. The signs cannot be obtained by this analysis. By definition, $2D_+$ is the hypotenuse of the triangle in which $\frac{1}{2}(J_{AX} - J_{BX})$ has been added to $\Delta\nu_{AB}$. When starting with a spin pattern, however, it cannot be

TABLE 4-4

Analysis of an ABX Pattern

I. Determine J_{AB} from the AB pattern. This separation occurs four times.

II. Determine $2D_+$ and $2D_-$ These cannot be distinguished from each other. Assume that the larger value is $2D_+$ The relative, but not the absolute, signs can be established later.

III. Construct two triangles having J_{AB} as bases and $2D_+$ as the hypotenuse of one and $2D_-$ as the hypotenuse of the other triangle.

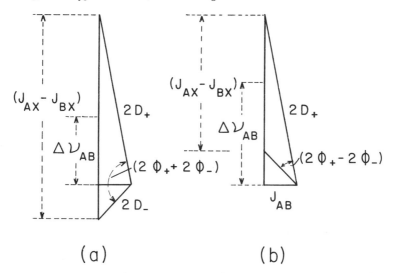

(a)   (b)

IV. Determine the angles $\frac{1}{2}(2\phi_+ + 2\phi_-)$ and $\frac{1}{2}(2\phi_+ - 2\phi_-)$. The ratio of the intensities of the medium to the strong peaks in the X pattern is equal to the square of the cosine of one of these angles. Select the correct angle. If the sum $\frac{1}{2}(2\phi_+ + 2\phi_-)$ gives the correct angle, then the two triangles are related as in (a). Otherwise they are related as in (b). It is not always possible to make this decision.

V. Determine $|\Delta\nu_{AB}|$ and $|J_{AX} - J_{BX}|$ by noting that in (a) $|J_{AX} - J_{BX}|$ is equal to the sum of the two vertical sides. In (b), $|J_{AX} - J_{BX}|$ is equal to the difference between the two sides. The halfway point on the combined vertical side in (a) or the point halfway between the vertical apexes in (b) is marked. The distance from this mark to the base is $|\Delta\nu_{AB}|$.

VI. From the X pattern find $|J_{AX} + J_{BX}|$, which is the separation between the two strongest peaks. The centers of the two AB patterns should be separated by $\frac{1}{2}|J_{AX} + J_{BX}|$ The medium peaks in the X pattern should be separated by $|2D_+ - 2D_-|$ The weakest peaks should be separated by $|2D_+ + 2D_-|$.

VII. From the values obtained in V and VI, $|J_{AX}| = \frac{1}{2}(|J_{AX} + J_{BX}| + |J_{AX} - J_{BX}|)$ and $|J_{BX}| = |J_{AX} + J_{BX}| - |J_{AX}|$.

VIII. The relative intensities of the peaks in the AB pattern may be calculated as indicated for the AB case in Figure 4-3.

determined which of the two measured distances is $|2D_+|$ and which is $|2D_-|$. This assignment could be made in Figure 4-5 because the pattern was calculated on the basis of assumed values for the various coupling constants and chemical shifts. For convenience, $|2D_+|$ will be taken as the larger of the two separations $|2D_+|$ and $|2D_-|$. It will be further assumed that J_{AB} is positive.

The magnitudes of the vertical sides of the triangles can be determined by construction or calculation since $|J_{AB}|$, $|2D_+|$, and $|2D_-|$ are all known. Addition and subtraction of these two vertical sides give two possible values for $|J_{AX} - J_{BX}|$. Both values must be considered at this point because the relative signs of $\Delta\nu_{AB}$ and $\frac{1}{2}(J_{AX} - J_{BX})$ are not known. This is the same as saying that it is not known whether the triangles are related as

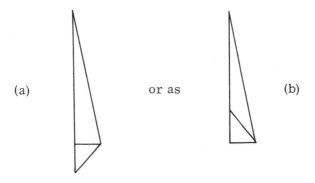

(a)   or as   (b)

Quite often the relationship of the two triangles to each other can be decided on the basis of the relative intensities of the peaks in the X portion of the pattern. When the triangles are related as in (b) above, the X portion will consist of only four peaks having approximately equal intensities. When the triangles are related as in (a), all six possible peaks may appear, in which case there will be two strong peaks, two medium peaks, and two weak peaks. The accuracy of the intensity measurements does not always permit a decision as to the relationship between the two triangles. This decision often establishes whether J_{AX} and J_{BX} have the same or opposite signs. The relative intensities of the X peaks can be dealt with quantitatively using the angles $2\phi_+$ and $2\phi_-$ (see the triangles in Figure 4-5). If one of the triangles is inverted because the two triangles are related as in (a) above, the corresponding angle is taken as negative. When the difference between one half of the angles $2\phi_+$ and $2\phi_-$ [that is, $\frac{1}{2}(2\phi_+ - 2\phi_-)$ or $(\phi_+ - \phi_-)$; see Figure 4-5] is small, two of the predicted peaks have a very low intensity.

For example, when $(\phi_+ - \phi_-)$ is 20°, the smallest peaks have intensities equal to 12% of the intensities of the strongest X peaks. Thus only four of the peaks may be observed. As $(\phi_+ - \phi_-)$ is made larger, all of the predicted six peaks appear. If the two strongest peaks in the X pattern are assigned an arbitrary integrated intensity of unity, the intensity of the medium peaks is $\cos^2(\phi_+ - \phi_-)$ and the intensity of the weaker peaks is $\sin^2(\phi_+ - \phi_-)$. Thus the combined intensity of the weaker and medium peaks is always equal to the intensity of the strongest peaks $(\cos^2\theta + \sin^2\theta = 1)$.

Once the relationship of the two triangles is established, the correct value of $|J_{AX} - J_{BX}|$ can be selected from the two possibilities determined above. The corresponding value for $|\Delta\nu_{AB}|$ can be calculated or determined by construction very simply. If the triangles are related as in (b) above, $\Delta\nu_{AB}$ is equal to one half the difference of the vertical sides of the two triangles. If the triangles are related as in (a) above, $|\Delta\nu_{AB}|$ is equal to one half the sum of the two vertical sides.

At this point $|J_{AX} - J_{BX}|$ and $\Delta\nu_{AB}$ are both known, provided, of course, that the difference in intensities of the X peaks permitted the choice between the possible values. From the X portion (lower section of Figure 4-5), the values of $|J_{AX} + J_{BX}|$ can be measured directly as the separation of the two strongest peaks. If there is difficulty in choosing the two strongest peaks, it will be necessary to examine the AB pattern. The centers of the two AB patterns are separated by $\frac{1}{2}|J_{AX} + J_{BX}|$. With both $|J_{AX} - J_{BX}|$ and $|J_{AX} + J_{BX}|$ known, the possible values of $|J_{AX}|$ and $|J_{BX}|$ can be calculated. Under favorable conditions, then, the quantities which can be determined from the ABX pattern are: $|J_{AB}|, |\nu_X|, |\Delta\nu_{AB}|, |J_{AB}|$, and $|J_{BX}|$. The relative signs of J_{AX} and J_{BX} can also be determined. None of the absolute signs of the three coupling constants can be established. For convenience, the sign of J_{AB} is usually taken as positive.

With this background discussion of the ABX system, the following important concepts will be considered:

1. Even if J_{AX} or J_{BX} is zero, the X pattern can still consist of as many as six peaks rather than the intuitively expected two peaks.

2. The X pattern will consist of a triplet under two circumstances [48]:

    a. if $J_{AX} = J_{BX}$; or,
    b. if J_{AB} is large compared to $\Delta\nu_{AB} \pm \frac{1}{2}(J_{AX} - J_{BX})$.

From the X pattern in the lower section of Figure 4-5 it will be seen that there are two ways in which the peaks could be superimposed to give a doublet. Either $|J_{AX} + J_{BX}|$ could be equal to $|2D_+ - 2D_-|$ or to $|2D_+ + 2D_-|$. The intensity of the remaining two peaks must, at the same time, be approximately equal to zero. The graphical interpretation of these requirements is that $|J_{AX} + J_{BX}|$ be equal to either the sum or the difference of the hypotenuse of the upper triangle and the hypotenuse of the lower triangle. The low intensity of the two remaining peaks could be brought about by having $\frac{1}{2}(2\phi_+ - 2\phi_-)$ either very small [for peaks separated by $(2D_+ + 2D_-)$] or close to 90° [for peaks separated by $(2D_+ - 2D_-)$]. This would make either $\sin^2(\phi_+ - \phi_-)$ or $\cos^2(\phi_+ - \phi_-)$ very small and the other value approximately equal to unity. Many combinations of variables could undoubtedly be found which would bring about the required equality and also cause the extra peaks to have a very low intensity. Even if the pair of strong peaks did not exactly coincide, the separations might be too small to be detectable. Making either J_{AX} or J_{BX} equal to zero, however, would not necessarily cause the X pattern to consist of a doublet. This would simply mean that $|J_{AX}|$ or $|J_{BX}|$ would have to be equal to $|2D_+ - 2D_-|$ or $|2D_+ + 2D_-|$. A condition which would essentially bring about one or the other of these equalities and would at the same time ensure that the extra peaks were very small would be that $\Delta\nu_{AB}$ be at least three times J_{AB}. This would make the triangles tall and thin, so that J_{AX} (or J_{BX}) would be essentially equal to either the difference or sum of the hypotenuse of the one triangle and the hypotenuse of the other triangle. This will be recognized as a somewhat less restrictive form of the first-order requirement that $\Delta\nu_{AB}$ be at least six times J_{AB}.

In an AMX system, then, if J_{AX} were zero and J_{MX} were some value other than zero, the X proton signal would always be a doublet with a separation equal to J_{MX}. On the other hand, in an ABX system, if J_{AX} were zero, the X signal could consist of as many as six peaks. The two strongest peaks would still be separated by J_{BX} (or actually by $|J_{AX} + J_{BX}|$ where $J_{AX} = 0$; see Figure 4-5), but none of the other separations would correspond to coupling constants. The term "virtual coupling" has been applied to systems of this type in which extra peaks are seen even though one of the coupling constants is zero [56]. The extra peaks arise because the system is higher-order. An incorrect interpretation of the extra splittings as being due to additional coupling could result from the incorrect first-order treatment of this system which does not conform to first-order conditions.

The important point here is that the first-order multiplicity rules must not be applied to higher-order systems. In general,

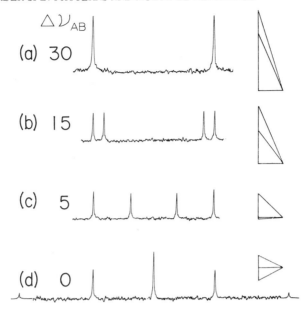

Fig. 4-6. Changes brought about in the calculated X pattern of an ABX system by decreasing $\Delta \nu_{AB}$. In each of these patterns the following values were held constant: $J_{AB} = 11$ cps, $J_{AX} = 0$, and $J_{BX} = 11$.

neither the coupling constants nor the chemical shifts can be extracted from these higher-order patterns by simple inspection.

A specific example which illustrates the problems which can arise in an ABX type is given in Figure 4-6. The patterns shown are those calculated for the X proton of an ABX system in which $J_{AX} = 0$, $J_{BX} = 11$, and $J_{AB} = 11$. The patterns show the effect of decreasing the chemical shift between the A and B protons. If $\Delta \nu_{AB}$ is 30 cps or greater, the X pattern will consist of two peaks separated by J_{BX} as shown in Figure 4-6a. If $\Delta \nu_{AB}$ is decreased to 15 cps (Figure 4-6b), each of the two peaks will break up into doublets separated by 1.1 cps. At the point at which $\Delta \nu_{AB}$ is 5 cps (Figure 4-6c), the members of the two doublets are separated by 3.4 cps. As $\Delta \nu_{AB}$ goes to zero, the two center peaks merge to give a single peak (Figure 4-6d). The final pattern appears as a triplet. Closer examination of the pattern shows two extra peaks separated by 24.6 cps and having intensities equal to 20% of the intensities of the outer peaks of the triplet. If the presence of the two weak outer peaks were overlooked, the pattern in Figure 4-6d might be mistakenly interpreted as a single triplet due to equal coupling with two other protons. The actual case is that X is coupled with only one other proton.

In general, a higher-order pattern cannot be completely analyzed when only part of the total pattern can be seen. Any conclusions drawn exclusively from the X portion of an ABX system, for example, must be accepted with considerable caution. Different combinations of constants may give indistinguishable patterns.

Frequently, a part of a molecule is considered as a spin system. For example, if three protons were on consecutive carbon atoms

$$\overset{\overset{\displaystyle OH}{|}}{\text{such as } -CH_A-CH_B-CH_X-}$$, the hydroxyl end of the system might be considered (with fast exchange or after D_2O exchange) to be an AX type. Since J_{AX} is usually almost zero, the X pattern might be expected to be a simple doublet separated by J_{BX}. As shown above, this could be depended on only if $\Delta\nu_{AB}$ were at least 3 J_{AB}.

The general rule for working with unisolated systems of protons is as follows: In a system of the type $A_M B_N X_P$ where $J_{AX}=0$, the X_P group can be considered as part of a $B_N X_P$ system only if $\Delta\nu_{AB}$ is greater than three times J_{AB} [58]. This rule is based on proton systems only and must be modified if coupling constants greater than 15 cps are involved. If $\Delta\nu_{AB}$ is less than three times J_{AB}, the system must be dealt with in its entirety.

Fortunately, the same electronic influences which shift one group of protons into a relatively clear part of the spectrum also usually cause a reasonable chemical-shift difference in the nearby groups of protons.

When J_{AX} is equal to J_{BX}, $\frac{1}{2}(J_{AX} - J_{BX})$ is equal to zero. From Figure 4-5 it will be seen that the vertical sides of the two triangles will then both become equal to $\Delta\nu_{AB}$. Since the bases of the triangles are also equal (J_{AB}), the triangles are then congruent. This in turn means that $2D_+$ is equal to $2D_-$ and that $(2D_+ - 2D_-)$ is equal to zero. Two of the peaks in the X portion coalesce to a single peak. At the same time $\frac{1}{2}(2\phi_+ - 2\phi_-)$ becomes equal to zero and so the intensities of the peaks separated by $(2D_+ + 2D_-)$ become equal to zero. The final result is the expected $1:2:1$ triplet for the X signal. As far as the X proton is concerned, this is a first-order coupling. The X proton is coupled equally ($J_{AX} = J_{BX}$) to two other protons which are separated ($\Delta\nu_{AX}$ and $\Delta\nu_{BX}$) from the X proton by at least six times the corresponding coupling constant (J_{AX} or J_{BX}). The corresponding AB pattern is not first-order because at least one of the first-order conditions ($\Delta\nu_{AB} > 6J_{AB}$) is not met.

The X pattern of an ABX system will also appear as a triplet if $\Delta \nu_{AB}$ is small compared to J_{AB} [48]. An example of this was seen in Figure 4-6d. The outer peaks of the "triplet" were separated by ($J_{AX} + J_{BX}$). The center peak was due to two peaks which coincided because $2D_+$ was equal to $2D_-$, making the difference ($2D_+ - 2D_-$) equal to zero. The actual requirement for causing the X pattern to appear as a triplet is that $\Delta \nu_{AB} \pm \frac{1}{2}(J_{AX} - J_{BX})$ be small compared to J_{AB}. This is the extreme departure from the first-order condition which requires that $\Delta \nu_{AB}$ be at least six times J_{AB}. The graphical interpretation is that the two triangles become long (horizontally) and thin. This makes the hypotenuse of one of the triangles become essentially equal to the hypotenuse of the other triangle, so that ($2D_+ - 2D_-$) is essentially zero. [If $\Delta \nu_{AB}$ is zero, then ($2D_+ - 2D_-$) is zero regardless of the value of J_{AB}.]

When $\Delta \nu_{AB} \pm \frac{1}{2}(J_{AX} - J_{BX})$ is large compared to J_{AB}, the appearance of the X signal as a triplet leads to the statement of a general rule [71]. If the differences in chemical shifts among a group of different nuclei are small compared to the coupling constants among that group of nuclei, then the pattern for the group of different nuclei will appear as if the nuclei are equivalent and are equally coupled to neighboring protons. Thus in an ABX system, if $\Delta \nu_{AB}$ is very small compared to J_{AB}, the spin pattern will resemble the pattern of an A_2X system. An ABC system in which $\Delta \nu_{AB}$ is small compared to J_{AB} will appear as an A_2B type. This is intuitively unexpected and represents an extreme departure from the first-order condition that $\Delta \nu_{AB}$ be at least six times J_{AB}.

Actually, the multiplicity rule which states that equivalent nuclei cannot interact to cause observable multiplicity can be considered empirically as a limiting case of the general rule stated above. It was shown in Figure 4-1 that, as the ratio of $\Delta \nu$ to J is decreased, the pattern eventually appears as a single peak. The value of $\Delta \nu$ at which this occurs is larger for more strongly coupled protons. For example, the signals due to two different and uncoupled protons can be easily distinguished from each other if they are separated by 1 cps. If these same protons were coupled with each other with $J_{AB} = 10$ cps, the calculation shows that the two strong inner peaks would be separated by only 0.5 cps when $\Delta \nu_{AB}$ was 3.2 cps. The outer peaks of the AB pattern would not be easily observed since the intensities of these peaks would be only 2.5% of the intensities of the inner peaks.

A rather remote possibility which would also lead to a triplet for the X pattern would be for J_{AX} to equal $-J_{BX}$ and at the same time for $\Delta\nu_{AB}$ to be large compared to J_{AB}. The center peaks in the X portion of Figure 4-5 would coalesce ($J_{AX} + J_{BX} = 0$) and the outer peaks would be of low intensity.

Both of the two general rules given above involve departures from the first-order condition that $\Delta\nu$ be at least six times J. As an AMX system is changed into an ABX type, the X pattern is no longer dependent only on J_{AX} and J_{BX}, but also on $\Delta\nu_{AB}$ and J_{AB}. When $\Delta\nu_{AB}$ become less than three times J_{AB}, the X pattern can consist of four (or six) peaks even though J_{AX} is equal to zero. As $\Delta\nu_{AB}$ becomes much less than J_{AB}, the extra peaks merge in the center of the pattern. The result is a triplet. The effect of decreasing the ratio of $\Delta\nu_{AB}$ to J_{AB} has been to transform the AMX into an ABX and finally into an A_2X type.

In an ABX system, when $\Delta\nu_{AX}$ and $\Delta\nu_{BX}$ are decreased to less than six times J_{AX} or J_{BX}, the system becomes an ABC type. The same higher-order effects noted earlier can be seen as $\Delta\nu_{AX}$ and $\Delta\nu_{BX}$ are decreased relative to J_{AX} and J_{BX}. As the ABX type is changed into an ABC type, the intensities of the peaks toward the center of the pattern increase at the expense of the peaks away from the center of the pattern. Thus the intensities of the peaks in both the X pattern and the AB pattern appear to slant toward the other part of the pattern. A new peak appears to give a maximum of 15 peaks (one more than for the ABX system). Even J_{AB}, which can be measured directly in the ABX pattern, cannot be determined by simple inspection of the ABC pattern.

Coupling constants can be either positive or negative. The appearance of first-order patterns is unaffected by a change in the sign of the coupling constant or by a difference in the signs if two different coupling constants are involved. The patterns due to AB and the three-spin systems AB_2, $A'B_2'$, and $A'X_2'$ reveal nothing about relative signs of coupling constants because either only one coupling constant is involved (AB and AB_2) or the pattern is a function only of the ratio of $\Delta\nu_{AB(or\ AX)}$ to $J_{AB(or\ AX)\ average}$. In other higher-order systems a change in the relative signs of coupling constants may cause a significant and predictable change in the appearance of the patterns. For example, for the ABX system it was pointed out that the choice of the correct values for J_{AX}, J_{BX}, and $\Delta\nu_{AB}$ can often be made by considering the relative intensities of the peaks in the X portion.

It has been proposed that, in general, for saturated systems the sign of the coupling constant alternates with each additional bond [37]. This means that if the sign of the coupling constant between protons on the same carbon atom is taken as negative, then the sign of the coupling constant between protons on adjacent carbon atoms is positive. The most direct method of determining relative signs is by the double-resonance technique, which will be discussed later in this chapter.

Considerable complexity can arise among some of the four-spin systems. Those which give first-order spectra are the A_4, A_3X, and A_2X_2. The higher-order patterns which will be discussed are the AB_3, $A_2'X_2'$, A_2B_2, and $A_2'B_2'$.

The AB_3 system frequently occurs in the form CH_3-CH-. If this system is not isolated from other protons, the higher-order effects noted in the ABX case must be considered [57]. Methylcyclohexane, for example, might be said to illustrate an A_3B system. In reality, however, this is an $A_3BC_2D_2E_2F_2GH$ system! As far as the appearance of the methyl pattern is concerned, it would be sufficient to call this an $A_3BC_2D_2$ type. This distortion of the A_3B doublet of methylcyclohexane can be seen in Figure 4-7. Some of the distortion is due simply to the departure of the A_3B portion from the first-order condition that $\Delta\nu_{AB}$ be at least six times J_{AB}. Further distortion occurs because (1) the B proton is coupled strongly to the adjacent methylene protons, and (2) the difference in chemical shift between the B proton and the adjacent methylene protons is not very large. The separation of the distorted doublet is not equal to the coupling constant between the methyl protons and adjacent methine proton [57].

The four-spin systems which involve two protons of each of two types (A_2B_2, $A_2'B_2'$, A_2X_2, and $A_2'X_2'$) all give patterns which are symmetrical about the midpoints of the patterns. Furthermore, the $A_2'X_2'$ and A_2X_2 patterns are also symmetrical about the midpoints of each half of the patterns. Examples of the A_2B_2 and $A_2'B_2'$ types are illustrated in Figure 4-8. Figure 4-1a shows an A_2X_2 type. These spin systems are easily identified by the symmetry of their patterns. The A_2X_2 type gives a first-order pattern. The remaining three types, A_2B_2, $A_2'B_2'$, and $A_2'X_2'$, occur frequently in a limited number of distinctly different organic groups. The distortions seen in these higher-order systems are useful in identifying the arrangement of protons in the groups. Typical patterns are indicated in Figure 4-8. Inspection of these four-spin patterns reveals that in the aliphatic series the outer peak is the first to break up into further peaks. For the $-CH_2-CH_2$ system in the cyclic series, the first extra peak occurs between the center and inner peaks. The possible involvement of two different coupling constants between the

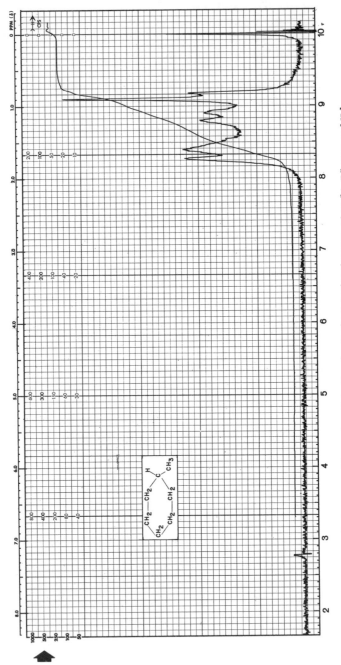

Fig. 4-7. Spectrum illustrating distortion due to the unisolated nature of an "A$_3$B" system [57].

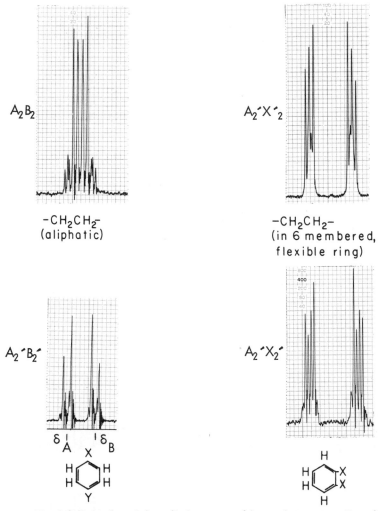

Fig. 4-8. Examples of four distinct types of four-spin systems. Note the symmetry about the center of each pattern. The approximate spin system is indicated. Neither the coupling constants nor the chemical shifts can be extracted by simple inspection of the A_2B_2 and $A_2'B_2'$ patterns. The centers of each of the two bands in the $A_2'X_2'$ patterns can be taken as the approximate chemical shifts, but the coupling constants cannot be extracted by inspection. These patterns are reprinted from Varian Catalogs [7,8] by permission of the copyright holder, Varian Associates.

two groups in compounds of the type XCH_2CH_2Y was discussed on p. 80. The p-disubstituted phenyl system appears as a typical double AB pattern with small extra peaks at the base of each main peak. The pattern due to an o-disubstituted phenyl type can be distinguished from the cyclic $A_2'B_2'$ pattern by the extra peaks in the center of the pattern and the intensities of the middle peaks in each half of the pattern.

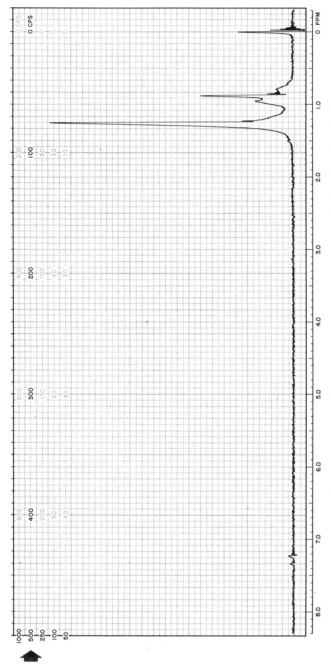

Fig. 4-9. Spectrum of N-octane illustrating the usual extreme in distortion seen in the methyl triplet of an ethyl group. Reproduced by permission of the copyright holder, Varian Associates.

The features which distinguish the four-spin systems from each other result from the similarities of the various coupling constants for each type.

One of the commonly encountered five-spin systems is the A_3B_2 resulting from an ethyl group. As $\Delta\nu_{AB}$ becomes less than six times J_{AB}, the expected distortion of intensities occurs and extra peaks appear. The small spacings of the extra peaks usually result in unresolved broad bands. The inner peak of the triplet becomes broader than the middle peak. The A_3 pattern may be further broadened if the system is not isolated. The usual extreme in distortion is seen in long-chain alkyl groups. An example is given in Figure 4-9.

Table 4-5 summarizes all the points which must be remembered in working with higher-order systems. The prediction of spin patterns for common systems is summarized in Table 4-3.

TABLE 4-5

Summary of Points Which Must be Remembered in Working with Higher-Order Systems

I.	Departure from one or both of the two first-order conditions causes (1) distortion of intensities (slanting towards other band), (2) appearance of extra peaks, and (3) unequal separations.
II.	Intuitive analysis of higher-order patterns is not possible.
III.	Chemical shifts and coupling constants cannot be extracted from higher-order patterns by simple inspection.
IV.	Unequal coupling constants are of concern in all higher-order systems except in $A_2'B'$, and $A_2'X'$.
V.	A change in relative signs of coupling constants may cause changes in the patterns of higher-order systems involving three or more nuclei.
VI.	For a system of the type $A_M B_N X_P$, the X_P group can be considered as part of a $B_N X_P$ system only if $\Delta\nu_{AB}$ is at least three times J_{AB}. Otherwise the entire system must be dealt with.
VII.	If the difference in chemical shifts among a group of nonequivalent nuclei is small compared to the coupling constants among the nuclei, the nonequivalent nuclei may behave like a group of equivalent nuclei which are equally coupled to neighboring nuclei.

Spin–spin coupling patterns can be conveniently placed in the following three categories:

1. First-order patterns which are readily identified by inspection,
2. Higher-order patterns which can be identified by their resemblance to first-order splittings,
3. Higher-order patterns which are sufficiently complex to require considerable work to firmly establish their identity.

Multiplets due to types such as AX, AMX, ABX, $ABCX_2$, and $A_2'B_2'X_2'$, in which two groups of very dissimilar nuclei are coupled, can be grouped as a fourth class of patterns. The coupling constants may be unequal. In all of these systems, application of a technique known as "double resonance" or "double irradiation" [12] will remove the couplings to the X nuclei and reduce the patterns to those expected of the types A, AM, AB, ABC, and $A_2'B_2'$, respectively.

The usual spectrum, which is a single-irradiation experiment, is obtained by applying a fixed radio-frequency signal to the sample while the magnetic field strength is gradually increased. Nuclei in different environments absorb radio-frequency energy at different magnetic field strengths. In the multiple-irradiation experiment, two or more fixed radio-frequency signals are used. If the difference in the frequencies is properly adjusted, two or more groups of protons will absorb energy at the same magnetic field strength. In order to observe complete decoupling of nucleus X from A in an AX system, the radio-frequency signal used for irradiating X must be very large while that used for A must be very small. Once the experimental details are arranged, only the difference in the two radio frequencies needs to be adjusted. Table 4-6 summarizes these experimental requirements.

A notation suggested [12] for indicating spin–spin decoupling by double irradiation is to first write the nucleus which is observed, followed in brackets by the nucleus from which it is decoupled. Thus, for an AX system, if the A signal was observed while X was decoupled from A, the experiment would be written $A-\{X\}$. The same type experiment for an A_2B_2X case would be written $A_2B_2-\{X\}$.

Double resonance can be used to solve another commonly occurring problem. If part of a system which can be decoupled is hidden by other bands, the difference in chemical shifts can be determined by experimentally finding the difference in irradiating frequencies required to decouple the system.

TABLE 4-6

Some Conditions Required to Decouple Nucleus X (Lower Field Signal) from Nucleus A (Higher Field Signal) for an AX System by Double Resonance

I.	The two radio frequencies used must differ approximately by the difference in chemical shifts (X frequency higher).
II.	Power in the radio-frequency signal used for:
	A. nucleus X must be large,
	B. nucleus A must be small.

Complete decoupling of higher-order systems, such as the AB type, cannot be achieved by double irradiation, Actually, the patterns become more complex. A further limitation is imposed by the difficulty of keeping the two radio-frequency signals from interfering with each other as the difference in frequencies becomes small.

The spin decoupling of nuclei of different elements, such as $H^1-\{F^{19}\}$, is referred to as "heteronuclear double irradiation." If the experiment involves nuclei belonging to the same element, such as $H^1-\{H^1\}$, the process is "homonuclear." The large difference in the frequencies required for nuclei of different elements makes the experiment easier to perform. In cases such as $H^1-\{F^{19}\}$, $H^1-\{N^{14}\}$, and $H^1-\{P^{31}\}$, the chemical shift of the unobserved nucleus is also determined by the required difference in frequencies.

In theory, any number of decoupling frequencies could be used, but practical considerations restrict the usual experiments to double and triple spin decoupling.

Decoupling by double irradiation gives the same result as decoupling by promoting rapid exchange of the hydroxyl proton in a group such as $-\overset{|}{C}HOH$ (see, for example, Figure 3-4). The nature of the experiments differ [12], however. Decoupling by promoting rapid exchange is brought about by averaging out the differences in environment of the proton attached to the carbon atom. Decoupling by double irradiation is achieved by regulating the relative orientations of the nuclear magnets.

The coupling between protons and deuterium nuclei is approximately $\frac{1}{7}$ of the corresponding proton–proton coupling. Replacement of protons by deuterium nuclei thus simplifies the remaining proton patterns. This simplification by isotopic substitution has been widely used. By double irradiation ($H^1 - \{H^2\}$), the pattern can be simplified even further because the remaining small couplings to the deuterium nuclei are removed.

Although the organic chemist is usually content simply to identify the more complex spin patterns, there are occasions when he may be interested in extracting as much information as possible. The details for working with AB and ABX systems were given earlier. For these and other fairly simple systems, the first step is a visual comparison of the observed multiplet with theoretically calculated patterns such as those published by Wiberg and Nist [6] and Corio [5]. When a reasonably close match is found, the scale of the calculated spectra can be adjusted and the pattern redrawn and compared directly with the observed pattern. The coupling constants and chemical shifts can then be obtained using the data given with the theoretical multiplet. The constants can then be adjusted by trial and error and the pattern recalculated using tables and a computer until any desired agreement is reached. Unfortunately, a very good fit may be obtained with incorrect coupling constants and chemical shifts. Part of this problem is solved by basing the possible values of the coupling constants on experience with similar, but simpler, systems. Computer programs are available for calculating patterns due to as many as nine nuclei. Programs and equipment are also available which print out the results as smooth curves which appear just as the actual patterns appear in the spectrum. The only input data required are the coupling constants and the chemical shifts. A systematic approach to the interpretation of spin patterns is outlined in Chapter 6.

A number of special problems can complicate the interpretation of spectra. These are discussed in the next chapter, along with techniques which can be employed to help solve these difficulties.

Chapter 5

Solutions of Problems Which Complicate the Interpretation of Spectra

Problems which frequently hamper the interpretation of spectra are outlined in Table 5-1. The most serious difficulties result from oversight of the possible problems.

Unsuspected contamination of the sample can cause considerable trouble. Two different types of problems may arise. First, if the impurity is present in sufficient quantity, any proton signals due to the impurity may be mistakenly assumed to be due to protons in the compound. Second, the signals due to impurities may accidentally appear to be components of spin multiplets.

In particular, the solvents used in the preparation and purification of the compound should be kept in mind. Many solvates are easily identified by their absorption peaks. Their composition may be determined using the integration curve. Solvents used in NMR frequently have proton-containing impurities. Usually those solvents which have completely deuterated methyl groups, such as CD_3COOD, contain a small amount of the compound containing only two deuterium atoms, such as CHD_2COOD. The resulting spectrum then shows a quintet of peaks ($2N + 1$; see Figure 3-6) centered at the position expected for the corresponding CH_3- group. These impurities are indicated in the solvent chart (Table 2-1) in Chapter 2. The peaks are separated by J_{H-D}, which is about 2.3 cps (0.154 times that expected for the corresponding J_{H-H}). Occasionally, an impurity may be washed out during D_2O exchange.

The sensitivity of proton signals to slight changes in molecular environments of the protons makes NMR a useful technique for detecting the presence of even closely related derivatives. Signals due to protons in isolated methyl groups are particularly suitable for this purpose. The presence of an impurity in which a sharp methyl peak is displaced by as much as 1 cps can cause detectable broadening of the peak. Although molecular changes in the immediate environment of a proton are most likely to cause the largest chemical shifts, changes in remote parts of the molecule can cause

TABLE 5-1

Problems Which Hamper the Interpretation
of Spectra

I.	Presence of several components:
	A. impurities: solvents, etc.
	B. molecular complexes
	C. slowly interconverting forms
	1. restricted rotation
	2. enol—keto
II.	Overlapping bands.
III.	Low resolution.
IV.	Unpredictable variations in coupling constants.
V.	Different possible interpretations.

easily detected displacements.   An illustration of the spectrum of
a mixture is given in Figure 5-1.

It can be seen that the two peaks at 57 cps (0.95 ppm), which are
due to the C-18 methyl protons in the two different compounds,
could be mistaken for a spin–spin doublet. The signals due to the
C-19 methyl protons of the two compounds coincide at 62 cps
(1.03 ppm).

Some aids which help to identify signals due to impurities are
listed in Table 5-2. As noted in Figure 5-1, the first indication of
an impurity which is usually noticed is the presence of an unex-
plainable peak or multiplet.   Examination of the integration curve
may show that less than a full proton is present in one or more regions.   In Figure 5-1, each portion of the AX pattern due to the olefinic protons represents less than a full proton.

The history of the sample or the spectrum may help to identify
the impurity.   Application of other physical methods, such as IR and
UV, may clearly establish the nature of the extra component.   Thin-
layer or gas chromatography may also be useful in determining the
number of impurities.   An examination of different preparations of

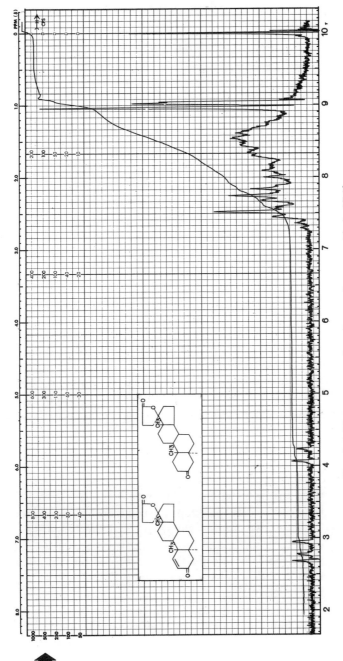

Fig. 5-1. The spectrum of a mixture of two steroids in CDCl₃.

TABLE 5-2

Aids for Identifying Signals Due to Protons in Impurities

I.	Presence of unexplainable single peaks or multiplets.
II.	Use of integration curve.
III.	Investigation of purity:
	A. history of sample and spectrum
	B. other physical methods (IR, UV, chromatography, etc.)
	C. spectra of different lots
	D. spectra of precursors and derivatives

the same material may be useful. The spectrum of a precursor or derivative may establish whether or not a peak is due to the desired compound. The spectra of these related compounds may also facilitate the interpretation of other aspects of the spectrum.

Molecular complexes which appear to be single compounds when examined by other methods frequently can be quickly recognized by their NMR spectra. Unrecognized as possible complexes, many of the differences in chemical shifts may be confused with spin–spin splittings.

Slowly interconverting forms give spectra in which each of the forms may show up separately. The peaks from the two forms may appear as spin multiplets. A survey of common types of slowly interconverting forms is given in Chapter 3.

Often the crux of the problem introduced by impurities is the distinction between two signals which are separated because of a difference in chemical shifts and those separated because of spin-spin splittings. The most direct answer is obtained by determining the spectrum at a different, preferably higher, frequency. Coupling constants are independent of the frequency used, while chemical shifts are proportional to the frequency. It was pointed out in Chapter 4 that coupling constants and chemical shifts can be determined by inspection only in a first-order spectrum. If a higher-order pattern is being considered, then the coupling constants and chemical shifts must be determined mathematically. The methods by which these calculations can be made were outlined in Chapter 4. The determination of the spectrum at a higher frequency will cause an increase in the ratio of $\Delta\nu$ to J. Multiplets due to higher-order spin systems will change towards first-order patterns. Comparatively few chemists have facilities available for determining spectra at two or more frequencies. Other methods must then be employed for distinguishing between chemical shifts and spin–spin splittings.

Chemical shifts are often sensitive to changes of solvent [60], while coupling constants are not appreciably affected. Thus a change of solvent may bring about a change in the ratio of $\Delta\nu$ to J. In some cases, this may help considerably in the analysis of the spin pattern. Pyridine is one of the best solvents for causing drastic changes in chemical shifts [61].

A change of solvent may eliminate another problem listed in Table 5-1; namely, the presence of overlapping bands. The clear separation of peaks is particularly helpful if an accurate comparison of the integrations of the peaks is desired.

The spectrum of the mixture of steroids used in Figure 5-1 is repeated in Figure 5-2 using pyridine as a solvent. Note that the different methyl proton signals now appear as a single peak. The peak at 278 cps (4.64 ppm) is due to water.

In the examination of small amounts of compounds, or compounds having low solubilities, the lack of good resolution may seriously handicap the interpretation of the spectra. Ways of increasing the signal-to-noise ratio are given in Chapter 2.

Some spin patterns are too complex to be resolved on currently available instruments. The spectra of closely related compounds can often be of tremendous help in the identification of these unresolved bands. If the protons involved are only remotely related to any changes in the molecule, the band will remain approximately the same width and shape and be in about the same position. The width of the band is controlled primarily by the sum of the coupling constants of the protons to neighboring protons. Higher-order effects can cause an increase or decrease in the width. The outer peaks may be so weak that they are lost in the noise. In order to overcome the difficulty of defining the limits of a broad band, the width at half-height can be used, for example, to determine whether steroidal secondary hydroxyl groups of the types $-CH_2CHOHCH-$ and $-CH_2-CHOH-CH_2-$ are axial or equatorial [62]. The width at

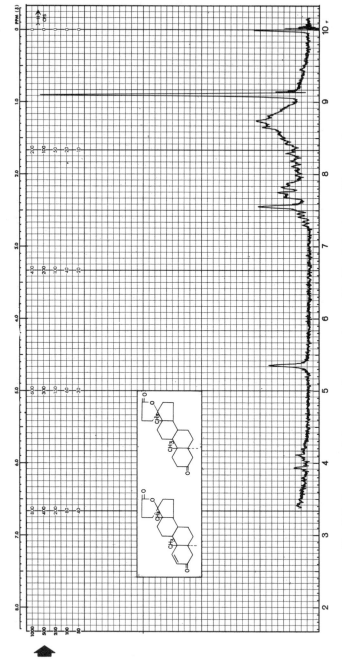

Fig. 5-2. Spectrum of the same mixture as that used in Fig. 5-1 using pyridine as solvent.

half-height of the band due to the proton on the same carbon atom as the hydroxyl group is less than 12 cps if the hydroxyl is axial, greater than 16 cps if the hydroxyl is equatorial. The width of these bands is controlled by the sum of the coupling constants of the proton to other protons. The axial hydroxyl has an equatorial proton attached to the carbon. This equatorial proton is coupled about equally (3.5 to 4 cps) to three or four other protons. The total width of the band then should be about 10.5 cps $(3 \cdot 3.5)$ to 14 cps $(4 \cdot 3.5)$. In the same manner, the equatorial hydroxyl has an axial proton attached to the same carbon atom. There are one or two axial–axial couplings and one or two axial–equatorial couplings. The width of the band then should be approximately 17 cps $(1 \cdot 10 + 2 \cdot 3.5)$ to 27 cps $(2 \cdot 10 + 2 \cdot 3.5)$. The complexities introduced by higher-order effects make it necessary to deal with these broad bands on a semiempirical basis.

Much remains to be learned about coupling constants. An open and cautious mind is the best guard against the many pitfalls. In general, the possibility of proton–proton coupling through four formally single bonds should be anticipated (1) when one or more of the intervening carbon atoms is doubly bonded to another group (such

$$
\overset{\displaystyle O}{\underset{\displaystyle \|}{}}
$$

as H—C—C—C—H) or (2) when the bonds to the protons appear as part of a W-shaped arrangement. Possible coupling through more than four bonds (such as H—C—C=C—C—H) should be considered whenever at least one of the intervening linkages is a multiple bond.

Spin–spin patterns calculated for more complex systems may agree closely with the observed spectrum and yet be determined using the wrong parameters.

When the NMR spectrum is used to help choose among a number of structures, it is essential that all of the possible structures be considered. The overlooked structure may fit the spectrum equally as well as the structure selected.

Even with all of the facts at hand, the interpretation of spectra may seem unnecessarily complex if a systematic approach is not employed. A detailed approach is suggested in the next chapter.

Chapter 6

A Systematic Approach to the Interpretation
of NMR Spectra

The essential features of previous chapters will now be referred to briefly in an outlined method of interpreting spectra. As an illustration, it will be assumed that a new product has been obtained from a known starting material. It will also be assumed that suitable tests have been made to determine purity, such as melting-point range, vapor-phase, and thin-layer chromatography.

The following spectra should be determined:

1. 1000-500 cps.
2. 500-0 cps.
3. Integration over 500-0 cps and any other regions containing absorption bands.

If the sample is suspected of having one or more protons attached to oxygen, nitrogen, or sulfur, then the sample should be exchanged with D_2O, and the spectra and integration should be repeated using the first spectra as a guide to the regions which should be examined. (A detailed procedure for the operation of the Varian A-60 is given in the appendix.)

The elemental composition of the sample and all other available physical measurements (IR, UV, and mass spectrum) should be determined.

The scheme recommended for interpreting the NMR spectrum starts with the simplest considerations, proceeds to the most complex, and then returns to the aspects of intermediate complexity. The features of the spectrum are examined in the following order:

1. General aspects
2. Integration curve
3. Signals due to exchangeable protons
4. Single, sharp peaks
5. First-order patterns
6. Unresolved bands
7. Resolved higher-order patterns

This analysis is followed by

    8. Correlation of the NMR spectrum with other data

    9. Overall review

The step-by-step program is given on the following pages.

OUTLINE OF PROCEDURE FOR INTERPRETING SPECTRA

1. Check the TMS signal for position, symmetry, and ringing. Correct the entire spectum for any error in the position of the TMS signal. Repeat the spectrum if the signal is not symmetrical or if inadequate ringing is observed.

Establish which regions are covered in the spectra; note the solvent employed.

Mark known impurities such as $CHCl_3$, CHD_2COOD, and HOD in exchange spectrum.

2. Calculate the relative numbers of protons represented by each band using one of the methods presented in Chapter 2. Make allowances for impurities.

3. Mark signals due to exchangeable protons by comparing the spectra obtained before and after the D_2O exchange. Make tentative identification from number, position, peak shape, multiplicity, elemental analysis and IR spectrum.

4. Identify single sharp peaks such as those due to CH_3O-,

$$CH_3-\overset{\displaystyle |}{\underset{\displaystyle |}{N}}-, \quad CH_3\overset{\displaystyle |}{\underset{\displaystyle |}{C}}-, \quad \text{and} \quad -\overset{\displaystyle |}{\underset{\displaystyle |}{C}}-C\equiv CH.$$

Remember the possibility of

accidental equivalence of protons. Check for proper relative numbers from the integration.

5. Pick out first-order multiplets by (1) their symmetry about the center of each portion of the pattern, (2) the relative intensities of the peaks (1:1, 1:2:1, etc.), and (3) the equal separations of the peaks. All separations between peaks in each portion of the pattern must be equal to the coupling constant (mutual coupling). Closer examination should reveal that in these first-order multiplets the peaks on the side toward the other part of the pattern are more intense than the corresponding peaks on the side away from the other part of the pattern. Furthermore, each multiplet in the pattern must be distorted to the same extent.

Consider possible couplings to nuclei other than protons if N^{14}, F^{19}, or P^{31} are present (Table 3-3, p. 61).

Assign $N-1$ protons to the group causing the splitting. Check the assignment of relative numbers against those obtained using the integration curve.

Connect groups in the manner dictated by the number of protons in the groups and by the magnitudes of the coupling constants (Tables 2-7a to 2-7d, pp. 37-40).

Propose environments for the protons in the first-order multiplets using other physical measurements, correlation charts (Figure 2-4, p. 16), or spectra of known compounds (see Table 2-6, p. 25).

Recall conditions which may produce unexpected nonequivalence of protons (Table 3-7, p. 74).

Consider the possibility of slowly interconverting forms, especially if nitrogen is present (Table 3-6, p. 68).

6. Identify broad unresolved bands. If possible, compare appearance, width at half-height, and position with similar bands of closely related model compounds.

7. Proceed to identify higher-order patterns (Tables 4-3, p. 82, and 6-1, p. 116). Note any gross resemblance to first-order patterns. Look for symmetry about center of pattern or portion of pattern. Check against relative numbers of protons in groups. Compare with expected higher-order effects (Table 4-5, p. 101). Consider the possibility of overlapping bands and presence of impurities. Repeat the spectra in a different solvent, if necessary, to change the chemical shifts or to help separate bands. Use multiple resonance to simplify patterns such as the A_2BX type.

Assign the nature of the four-spin types A_2B_2, $A_2'B_2'$, and $A_2'X_2'$ from examples in Figure 4-8, p. 99.

Assign molecular environments for the protons.

Check questionable assignments using model compounds or published calculated spectra. If desirable, calculate the pattern using successive estimates of the various parameters until a fit is obtained. Previous experience must be used as a guide in the selection of reasonable coupling constants and chemical shifts. A calculated pattern may be within the experimental error and still be based on incorrect parameters.

8. Combine the groupings suggested by this analysis with the functional groups determined by IR. Correlate the UV spectrum with any olefinic or aromatic proton signals. Propose a carbon skeleton from the chemistry, elemental analysis, and types of proton signals. The number of "rings" in a compound can be calculated [63] using the formula:

"Rings" = carbons + 1 - (hydrogens + halogens − trivalent nitrogens)/2

TABLE 6-1

Analysis of Spin Patterns from Their Characteristics

Number of protons from integration	Number of peaks n	Spacing of peaks	Relative intensities of peaks*	Possible systems†
1	2 to 9	Equal	Binomial	$\underline{A}X_{n-1}$
	2		Slanting	$\underline{A}B$
	3	Equal		$AB\underline{X}$: $J_{AX} = \lvert J_{AX} \rvert$ or J_{AB} $\gg \Delta\nu_{AB} + \frac{1}{2}(J_{AX} - J_{BX})$
	4 (or 4 + 2 weak)	Symmetrical about center of band		$AB\underline{X}$
	4	Unequal	Slanting	$A\underline{B}_2$
2	2 to 9	Equal	Binomial	$\underline{A}_2X_{n-1}$
	Complex: strong doublet + 2 quartets	Symmetrical about center of band; also mirror image of another 2-proton band		$\underline{A}'_2X'_2$
	Complex: as many as 12 peaks	Not symmetrical about center of band but mirror image of another 2-proton band		$\underline{A}_2B_2$ or $\underline{A}'_2B'_2$
	2 quartets (may be superimposed)	One spacing occurs 4 times	Like 2 AB systems	$A\underline{B}X$
	2 doublets (may have one other weak peak)	One doublet separated less than other	Slanting	$\underline{A}_2B$
3	2 or 3	Equal	Binomial	A_3X_{n-1}
	2 unresolved bands		Slanting	A_3B
	Complex	Complex	Complex	$AB\underline{C}$
	8 to 9 (4 peaks + 2 unequally spaced doublets + possible extra peak)	Unequal	Slanting	$A_2\underline{B}$
4	Complex	Symmetrical about center of band		$\underline{A}_2B_2$ or $\underline{A}'_2B'_2$
5	Complex	Complex	Complex	$\underline{A}_2B_2C$ (monosubstituted phenyl, for example)

*Binomial Intensities: 1:1
                        1:2:1
                        1:3:3:1
                        1:4:6:4:1
                        1:5:10:10:5:1
                        1:6:15:20:15:6:1
"Slanting" means general increase in intensities in direction of other part of pattern.
†The primed systems are those in which each A is *not* coupled equally to each B or X.

TABLE 6-2

Characteristics That Aid in
Analysis of Spin Patterns

I.	Symmetry of pattern
II.	Relative intensities of peaks
III.	Spacings of peaks
IV.	Number of protons represented
V.	Number of peaks
VI.	Correlation with other parts of pattern

"Ring" here is defined as any cyclic system. A double bond is counted as one ring, a triple bond as two rings. (Benzaldehyde has five "rings.") Propose structures that combine all of the data.

9. Review the spectrum for its correlation with the proposed structures. If possible, eliminate structures by using closely related models. Use multiple resonance, if available, to confirm which groups are coupled and to locate hidden protons.

It is to be noted that the first step in the analysis of any spin multiplet is the recognition of possible spin systems which could give rise to the pattern. There is no alternative to this initial identification by inspection. The characteristics which have been noted in this and previous chapters by which these identifications are made are summarized in Table 6-2. These characteristics form the basis for the correlations given in Table 6-1.

Equipped with this body of empirical facts, the student is ready to deduce a wealth of data from NMR spectra. Confidence can be derived only by the examination of many spectra. The spectra in the Varian catalogs [7,8] serve quite conveniently for this purpose. Experience with a series of closely related compounds can make the interpretation of spectra of related derivatives extremely easy. The student is also now in an excellent position to proceed with a study of the theory of NMR.

Appendix

Suggested Operating Procedure for the Varian A-60 NMR Spectrometer

by A. J. Damascus (G. D. Searle & Co.)

The exact procedure used in the tuning up and operation of a spectrometer varies greatly among the various laboratories. The following procedure is used at Searle.

A. Starting Instrument

1. Press black ON button to start instrument.
2. After cycling lamp goes out, place OPERATE-STANDBY switch in OPERATE. Adjust 5 Kcps on FREQUENCY METER.
3. Wait for magnet temperature to come to equilibrium (when the field meter stops drifting), then proceed to Part B.

B. Tuning Up Instrument

1. Readjust 5 Kcps if necessary.
2. Place water (HOMO-ADJUST) sample in a spinner turbine and insert into probe.
3. Open door to homogeneity and test controls.
4. Put homogeneity switch in ADJUST (spinner off).
5. Set X, Y, Z GRADIENTS (fine controls) and CURVATURE CONTROL to approximate center positions.
6. Switch positions:

RF FIELD	1.0
FILTER BANDWIDTH	4.0
SWEEP TIME	250
SWEEP OFFSET	000
SWEEP WIDTH	500
SPECTRUM AMPLITUDE	Varies with instrument, 1.0×1 is suggested as a start (keep pen on chart).
DETECTOR ZERO	Turned clockwise to limit.

PEN   AUTO
NORMAL SWEEP   OFF

7. Adjust X, Y, and Z COARSE CONTROLS, in that order, for maximum SIGNAL LEVEL METER deflection.

8. Adjust X and Z GRADIENTS for maximum pen deflection (more sensitive than meter deflection).

9. Start spinner. Synchronize to light (30 rps).

10. Adjust Y-COARSE for maximum pen deflection.

11. Adjust CURVATURE and Y-GRADIENT alternately for maximum pen deflection.

12. Repeat steps 8-11 until homogeneity is optimized (maximum pen deflection).

13. Put HOMOGENEITY switch in OPERATE. Move pen carriage to left end of chart by pressing fast sweep switch to the left. Press NORMAL SWEEP SWITCH to right and run spectrum of water. Adjust DETECTOR PHASE for a base line having same level on both sides of spectral line.

14. Remove HOMO-ADJUST sample and insert TMS sample.

15. Switch positions:

    RF FIELD   0.16 (see step No. 16)
    SPECTRUM AMPLITUDE 1.6×1 (varies with instrument)
    SWEEP TIME   250
    SWEEP OFFSET   000
    SWEEP WIDTH   500
    FILTER BANDWIDTH   4.0
    PEN   AUTO

16. Run through the TMS peak. Adjust Y-GRADIENT and curvature for maximum peak and ringing. Test curvature by scanning TMS peak in both directions. The two ringing patterns should be mirror images. If the patterns are not mirror images, the CURVATURE should be adjusted until this condition is met. With each CURVATURE adjustment, small adjustments should be made in the Y-GRADIENT to maximize the signal height and obtain symmetrical ringing. Touch up the DETECTOR PHASE if necessary. At this point, the RF level at which the instrument can usually be operated without danger of saturation can be determined. Repeatedly scan over the TMS peak, changing the RF level after each run. A RF level will be found at which the peak is of maximum amplitude without line broadening. This RF can now be used in the routine operation of the instrument. If saturation of an absorption signal is suspected, the area in question is rescanned at several different RF levels.

17. Adjust the SWEEP ZERO so that the TMS peak is on the zero line.

C. Running of Unknown Samples

1. Remove TMS sample and insert unknown sample.
2. Switch positions:

RF FIELD	0.16, or as determined in Part B, Step 16.
SPECTRUM AMPLITUDE	1.0×10
SWEEP TIME	250
SWEEP OFFSET	000
SWEEP WIDTH	500
FILTER BANDWIDTH	4.0
PEN	AUTO

3. Use the TMS in the sample to optimize peak and ringing, using Y-GRADIENT and CURVATURE. Adjust TMS to zero. Adjust base line to desired level with RECORDER ZERO.
4. Run a spectrum at SWEEP TIME of 100. Select tallest peak. Change SWEEP TIME to 250. Adjust the SPECTRUM AMPLITUDE until the tallest peak falls as high as possible without going off the chart. If the noise level is too high at this SPECTRUM AMPLITUDE, reduce the FILTER BAND-WIDTH. If the FILTER BANDWIDTH is below 2, the SWEEP TIME must be reduced to 500 and a check for saturation should be made.
5. Return pen carriage to left hand of chart and run spectrum.

D. Integration

1. After sample has been run, the following changes are made in switch positions:

RF FIELD	0.4-0.6
SPECTRUM AMPLITUDE	no change
SWEEP TIME	25
SWEEP WIDTH	no change
SWEEP OFFSET	no change
FILTER BANDWIDTH	no change
INTEGRAL AMPLITUDE	80
PEN POSITION	UP

(The integration is run as fast as possible and with a higher RF to improve stability and quality of integration. It is suggested that the integration be run at different RF levels to determine the saturation level as in Part B, Step 16.)

2. Using FAST SWEEP, bring pen over a region of absorption
    and let pen go to top of chart. Hold there for a few seconds
    (30 sec) to "warm up" integrator.
3. Bring pen to left end of chart and change SWEEP WIDTH
    to 1000.
4. Press INTEGRAL RESET to bring integrator and pen to
    zero. Hold down INTEGRAL RESET and adjust base line
    to desired level with RECORDER ZERO. Release INTE-
    GRAL RESET.
5. Press and release INTEGRAL RESET noting pen drift.
    Adjust the DETECTOR ZERO until any drift has been elimi-
    nated. Press and release the INTEGRAL RESET after
    each adjustment of the DETECTOR ZERO. Return the
    SWEEP WIDTH to 500.
6. Run through the integration with the pen up. Adjust the
    SPECTRUM AMPLITUDE to give the desired presentation.
    The DETECTOR ZERO may have to be readjusted when
    the SPECTRUM AMPLITUDE is changed.
7. Place pen in AUTO position. Return pen carriage to left
    end of chart and press INTEGRAL RESET. Press NORMAL
    SWEEP, and record integral.
8. When integration is complete, press INTEGRAL RESET
    and turn off integrator (INTEGRAL AMPLITUDE turned
    counterclockwise to limit).
9. Return RF and SWEEP TIME to normal settings. The in-
    strument is now ready for the next sample.

E. General Suggestions

1. Leave spinner air on at all times, even when instrument is
    not being used (this eliminates field shifts due to tempera-
    ture change when the air is turned on).
2. Change speeds only when pen carriage is running.
3. Phasing must be correct when integrating.

Calibration of the chemical shift scale using a mixture of compounds is described in reference 75. A convenient method of
determining the signal-to-noise ratio is given in reference 74.

References, Footnotes, and Selected Bibliography

Texts

1. J. A. Pople, W. G. Schneider, and H. J. Bernstein, "High-Resolution Nuclear Magnetic Resonance," McGraw-Hill, New York, 1959.
2. L. M. Jackman, "Applications of Nuclear Magnetic Resonance Spectroscopy in Organic Chemistry," Pergamon Press, New York, 1959.
3. J. D. Roberts, "Nuclear Magnetic Resonance," McGraw-Hill, New York, 1959.
4. J. D. Roberts, "An Introduction to Spin–Spin Splitting in High Resolution N.M.R. Spectra," W. A. Benjamin, New York, 1961.

Collections of Calculated Spin Patterns

5. P. L. Corio, Chem. Rev., 60:363 (1960).
6. K. B. Wiberg and B. J. Nist, "The Interpretation of NMR Spectra," W. A. Benjamin, New York, 1962.

Catalogs of Spectra

7. N. S. Bhacca, L. F. Johnson, and J. N. Shoolery, "NMR Spectra Catalog," Vol. 1, Varian Associates, Palo Alto, California, 1962.
8. N. S. Bhacca, D. P. Hollis, L. F. Johnson, and E. A. Pier, "NMR Spectra Catalog," Vol. 2, Varian Associates, Palo Alto, California, 1963.

References to Published Spectra

9. M. Gertrude Howell, Andrew S. Kende, and John S. Webb, [eds.], "Formula Index to NMR Literature Data," Vols. 1 and 2, Plenum Press, New York, 1965.

NMR of Nuclei Other Than H^1 and F^{19}

10. P. C. Lauterbur, "Nuclear Magnetic Resonance Spectra of Elements Other Than Hydrogen and Fluorine," in "Determination of Organic Structures by Physical Methods" (F. C. Nachod and W. D. Phillips, eds.), Vol. 2, Chap. 7, Academic Press, New York, 1962.

NMR of Solids

11. R. E. Richards, "Nuclear Resonance in Solids," in "Determination of Organic Structures by Physical Methods" (F. C. Nachod and W. D. Phillips, eds.), Vol. 2, Chap. 8, Academic Press, New York, 1962.

Multiple Resonance

12. J. D. Baldeschwieler and E. W. Randall, Chem. Rev., 63:81 (1963).

Chapters and Reviews

13. E. O. Bishop, Ann. Reports, 58:55 (1961).
14. W. D. Phillips, "High Resolution H^1 and F^{19} Magnetic Resonance Spectra of Organic Molecules," in "Determination of Organic Structures by Physical Methods" (F. C. Nachod and W. D. Phillips, eds.) Vol. 2, Chap. 6, Academic Press, New York, 1962.
15. J. B. Stothers, "Applications of Nuclear Magnetic Resonance Spectroscopy," in "Technique of Organic Chemistry" (A. Weissberger, ed.), Vol. XI (K. W. Bentley, ed.), Chap. IV, Interscience, New York, 1963.

Heterocyclic Compounds

16. R. F. M. White, "Nuclear Magnetic Resonance Spectra," in "Physical Methods in Heterocyclic Chemistry" (A. R. Katritzky, ed.), Vol. II, Chap. 9, Academic Press, New York, 1963.

Applications to Biological Problems

17. O. Jardetzky and C. D. Jardetzky, "Introduction to Magnetic Resonance Spectroscopy. Methods and Biochemical Applications," in "Methods of Biochemical Analysis" (D. Glick, ed.), Vol. IX, Interscience, New York, 1962.

Use of NMR in Conjunction with IR, UV, and Mass Spectra

18. R. M. Silverstein and G. Clayton Bassler, "Spectrometric Identification of Organic Compounds," Wiley, New York, 1963.

Determination of Structural Features in Steroids

19. N. S. Bhacca and D. H. Williams, "Applications of NMR Spectroscopy in Organic Chemistry. Illustrations from the Steroid Field," Holden-Day, Inc., San Francisco, 1964.

* * * * *

20. J. N. Shoolery and M. T. Rogers, J. Am. Chem. Soc., 80:5121 (1958).
21. C. E. Johnson, Jr., and F. A. Bovey, J. Chem. Physics, 29:1012 (1958).
22. Ref. 2, p. 116.
23. Varian Associates Tech. Information Bulletin, Vol. 2, No. 3.
24. Ref. 2, p. 59.
25. R. F. Zürcher, Helv. Chim. Acta, 46:2054 (1963).
26. H. S. Gutowsky and C. H. Holm, J. Chem. Phys., 25:1228 (1957).
27. Ref. 2, p. 85.
28. Ref. 1, p. 193.
29. D. P. Hollis, private communication.
30. A. C. Huitric, J. B. Carr, W. F. Trager, and B. J. Nist, Tetrahedron, 19:2145 (1963).
31. T. Takahashi, Tetrahedron Letters, 1964, No. 11, p. 565.
32. T. A. Crabb and R. C. Cookson, Tetrahedron Letters, 1964, No. 12, p. 679.
33. M. Karplus, J. Chem. Phys., 30:11 (1959).
34. M. Karplus, J. Am. Chem. Soc., 85:2870 (1963).
35. H. S. Gutowsky, M. Karplus, and D. M. Grant, J. Chem. Phys., 31:1278 (1959).
36. See the discussions given by (a) M. Karplus, J. Am. Chem. Soc., 84:2458 (1962), and (b) M. Barfield and D. M. Grant, J. Am. Chem. Soc., 85:1899 (1963).
37. J. D. Roberts, Angewandte Chemie, 2:53 (1963).
38. E. I. Snyder and J. D. Roberts, J. Am. Chem. Soc., 84:1582 (1962).
39. J. Meinwald and A. Lewis, J. Am. Chem. Soc., 83:2769 (1961).
40. (a) S. Sternhell, Rev. Pure and Appl. Chem., 14:15 (1964); (b) A. Rassat, C. W. Jefford, J. M. Lehn, and B. Waegell, Tetrahedron Letters, 233 (1964).

41. D. R. Davis, R. P. Lutz, and J. D. Roberts, J. Am. Chem. Soc., 83:246 (1961).
42. D. R. Davis and J. D. Roberts, J. Am. Chem. Soc., 84:2252 (1962).
43. Ref. 1, p. 98.
44. NMR Table published by Varian Associates, Palo Alto, California.
45. Ref. 1, Chap. 6.
46. Ref. 2, Chap. 6.
47. Ref. 1, p. 132.
48. Ref. 6, p. 21.
49. Ref. 2, p. 90.
50. See Ref. 1, p. 310.
51. Ref. 1, Chap. 13.
52. P. M. Nair and J. D. Roberts, J. Am. Chem. Soc., 79:4565 (1957).
53. Ref. 6, p. vi.
54. The general condition for possible nonequivalence of protons in a freely rotating methylene group is that "the plane bisecting the HCH angle and normal to the interproton axis is not a plane of symmetry of the molecule as a whole for any conformation." Quoted from E. O. Bishop, ref. 13, p. 67.
55. Ref. 4, p. 71.
56. J. I. Musher and E. J. Corey, Tetrahedron, 18:791 (1962).
57. F. A. L. Anet, Canad. J. Chem., 39:2262 (1961).
58. I am indebted to Dr. Frank A. L. Anet for his comments on the limiting ratio of $\Delta\nu$ to J for this rule. This limit is formulated conservatively.
59. Ref. 6, p. 27.
60. Ref. 1, Chap. 16.
61. G. Slomp and F. MacKellar, J. Am. Chem. Soc., 82:999 (1960).
62. Y. Kawazoa, Y. Sato, T. Okamoto, and K. Tsuda, Chem. and Pharmaceutical Bulletin, 11:328 (1963).
63. M. D. Soffer, Science, 127:880 (1958).
64. Ref. 6, p. 33.
65. Ref. 6, pp. 15, 16.
66. K. Tori and E. Kondo, Tetrahedron Letters, 1963, No. 10, p. 645.
67. Erno Mohacsi, J. Chem. Ed., 41:38 (1964).
68. A. D. Buckingham, Can. J. Chem., 38:300 (1960).
69. G. V. D. Tiers and R. I. Cook, J. Org. Chem., 26:2097 (1961).
70. O. L. Chapman and R. W. King, J. Am. Chem. Soc., 86:1256 (1964).
71. Ref. 6, p. 27.

72. Ref. 2, p. 71.

73. H. M. Fales and A. V. Robertson, Tetrahedron Letters, 1962, No. 3, p. 111.

74. G. Slomp, Abstracts of the 15th Mid-America Symposium on Spectroscopy, Chicago, Illinois, June, 1964, paper No. 105.

75. J. L. Jungnickel, Anal. Chem., 35:1985 (1963).

76. G. V. D. Tiers, J. Phys. Chem. 62:1151 (1958).

77. R. C. Hirst and D. M. Grant, J. Chem. Phys., 40:1909 (1964).

Glossary

The following terms are not explained in the text but are encountered in the literature. The definitions are intended to be descriptive rather than rigorous. Terms marked with an asterisk are listed separately in the glossary. The definitions have been formulated primarily from references 1, 2, and 3.

Anisotropy. Paired electrons in materials which are placed in a magnetic field tend to circulate in a direction such that the magnetic field generated by the motion of these electrons opposes the applied magnetic field. If the electrons tend to circulate more freely around one axis than another, the material is said to have an anisotropy of diamagnetic susceptibility.* The spatial effects of the functional groups in Figure 2-5 are a direct result of anisotropy.

Audio side band. Modulation of the magnetic field in which a sample is placed gives rise to new peaks which are symmetrically arranged about the original signals due to the sample. These new peaks are separated from the original signals by some multiple of the modulating signal. The modulating signal is usually in the audio range. With the aid of a calibrated signal generator, the separation of two peaks can be measured simply by determining the frequency required to make the side band of one peak coincide with the second peak. This is known as the "audiofrequency modulation" or "side-band" technique. Spinning side bands are formally similar to the side bands generated by this field modulation, but are caused by slow rotation of the sample tube in an inhomogeneous magnetic field. Spinning side bands are separated from the parent peak by multiples of the rate of rotation of the sample tube in revolutions per second.

Combination lines. Combination lines are peaks which result from the simultaneous change of spin states* of two or more nuclei. Peaks due to combination transitions (or "multiple quantum transitions") are not normally seen in first-order spectra but may appear under saturation* conditions.

Curvature. A synonym for the contour of the magnetic field strength of a magnet between the pole pieces. The contour is spoken of as "dome-shaped," "dish-shaped" or "flat." The absorption peaks will be distorted unless the curve is flat. The base line will be even on either side of the peak but the peaks will not be symmetrical about their centers. Adjustment of the curvature is part of the tune-up procedure. Rough adjustment is made by cycling* the magnet.

Cycling. The procedure of momentarily passing a larger than normal amount of current through an electromagnet and then decreasing the current to the normal operating level. The contour (or curvature*) of the magnetic field strength between the poles is made flatter by this operation. This is a rough adjustment. A finer adjustment is made electrically.

Detector phase control. The control used on the A-60 to remove the unwanted dispersion signal.* An incorrect adjustment causes the base line to be uneven on the sides of an absorption peak.

Diamagnetic screening. The shift of an NMR signal to higher magnetic field which is due to the circulation of paired electrons.

Dipole—dipole broadening. In the solid state or in a viscous liquid, the magnetic nuclei in a sample are mutually subjected to their own large magnetic fields. These fields are not averaged out (as they are in nonviscous solutions). The magnetic field strength thus varies from point to point, effectively producing an overall inhomogeneous magnetic field, which leads to broadening of the peaks. The signals due to the protons in the Varian nylon microcells are so broad, for example, that they can be ignored in the high-resolution spectrum.

Dispersion signal. The absorption signal normally plotted is accompanied by another signal called the "dispersion" or "u mode" signal. The dispersion curve goes below the base line, then through the base line at the point at which the absorption signal is at its maximum, and then above the base line. The dispersion signal is out of phase with the absorption signal and is eliminated by use of the "detector phase" control on the A-60. When the phasing is correct, absorption signals start and return to the same base line.

Gyromagnetic ratio (γ). A synonym for magnetogyric ratio.*

$\underline{I}$. See spin quantum number.*

Induction method. The induction (or "crossed coil") method, one of the two methods used for determining NMR spectra, utilizes two coils at right angles to each other. The signal picked up by one coil is due to the re-emission of the energy absorbed by the nuclei from the other coil. The other method, the single-coil method, which is used in the A-60, actually measures the change in the impedance of a single coil which is part of a tuned circuit.

Larmor frequency (ω). The angular velocity of nuclear precession.* Related to the magnetogyric ratio* γ and the applied magnetic field H by the equation

$$\omega = \gamma H$$

Lattice. The entire molecular environment, including the solvent, of a nucleus.

Longitudinal relaxation (T_1). The loss of nuclear spin* energy by transfer to the molecular environment. Also called spin-lattice relaxation. Brought about primarily by fluctuating local magnetic fields due (1) to other magnetic nuclei, (2) to unpaired electrons, or (3) variable screening of the applied magnetic field.

Magnetic induction (B). Magnetic flux produced per unit area measured perpendicular to the direction of the flux (unit: gauss).

Magnetic quantum number (nuclear) (m). A property of the nucleus which describes the discrete values of angular momentum (spin) which a given nucleus may exhibit. The magnetic quantum number may have the values

$$m = I, I - 1, I - 2, \ldots, -I + 1, -I$$

where I is the spin quantum number.* For protons, $I = \frac{1}{2}$ so that $m = +\frac{1}{2}, -\frac{1}{2}$. For $H^2, I = 1$ so that $m = +1, 0, -1$. The number of values which m may have is thus $(2I + 1)$.

Magnetization, intensity of (M). Magnetic moment produced per unit volume.

Magnetogyric ratio (γ). A nuclear property defined by

$$\gamma = \mu/I\hbar$$

where μ is the maximum observable component of the magnetic moment, I is the spin quantum number,* and $\hbar$ is the modified Planck constant ($\hbar = h/2\pi$). Also called gyromagnetic ratio.

Magneton, nuclear. A unit for the measurement of nuclear moments.

Moment, magnetic. The torque experienced by a magnet when it is at right angles to a uniform field of unit intensity. The value of the magnetic moment is equal to the product of the magnetic pole strength and the distance between the poles.

Nuclear side-band oscillator. A device in the A-60 by which the magnetic field strength and radio-frequency oscillator are "locked" together through a circuit which is tuned by the resonance signal of a water sample in the magnetic field.

Nuclear spin. The magnetic moment of protons and other magnetic nuclei results from the rotation of the charged nucleus. Non-magnetic nuclei do not possess nuclear spin.

Nuclear spin temperature. A term used to describe the distribution of nuclei in the various spin states. For protons (or any nuclei for which $I = \frac{1}{2}$), the spin temperature T_s is given by

$$T_s = \frac{2\mu H_0}{k \ln (n_+/n_-)}$$

where μ is the nuclear moment, H_0 is the magnetic field, k is Boltzmann's constant, and n_+ and n_- are the number of nuclei in the lower and upper states, respectively. The spin temperature is negative if the population in the upper state is greater than the population in the lower state.

Paramagnetic substances. See susceptibility.*

Pascal constants. Atomic contributions to the molar diamagnetic susceptibility.* The calculated molar diamagnetic susceptibility (χ_M) is equal to the sum of the atomic contributions. The volume susceptibility* is then obtained by the formula

$$\chi_V^{=} = \chi_M (d/\text{MW})$$

where MW is the molecular weight and d is the density of the substance.

Precession, nuclear. Those nuclei, such as H^1, which have magnetic moments behave like spinning magnets. When placed in an external magnetic field, these spinning magnets wobble about their axes of rotation. This wobbling is called precession and occurs at the Larmor frequency.*

Probe. The apparatus which fits between the pole faces of the magnet and which contains the sample chamber and necessary mechanical and electrical components for the irradiation of the sample with radio-frequency energy, detection of the nuclear signal, and the modulation and sweep of the magnetic field.

Pulse method. A special method of studying NMR using short bursts of radio-frequency energy at the resonance frequency. By the application of two or more properly timed bursts, a nuclear signal (spin echo) can be observed at a time during which no radio-frequency energy is being applied.

Quadrupole broadening. Increase in peak width due to decrease in relaxation time* caused by a nearby nucleus having a nonspherical charge distribution (quadrupole moment*). Signals due to protons attached to N^{14} can be broadened by this effect. (Note also the effects of spin–spin coupling and exchange in Tables 3-4 and 3-5, on pp. 66 and 67.)

Quadrupole coupling. The interaction between a nucleus having a nonspherical charge distribution and another magnetic nucleus. This gives rise to broadening of the signal due to the other magnetic nucleus. The coupling is least when the nucleus with the quadrupole moment is in an electrically symmetrical environment.

Quadrupole moment (electric) (Q). The moment which is a result of a nonspherical charge distribution on the nucleus. Nuclei which have $I > \frac{1}{2}$ may have electric quadrupole moments while those with $I = 0$ or $\frac{1}{2}$ do not. Common nuclei with quadrupole moments are H^2, B^{11}, and N^{14}.

Quadrupole relaxation. Loss of nuclear spin energy by interaction with the magnetic field produced by a nucleus having a non-spherical charge distribution (electric quadrupole moment*).

Quantum number, magnetic. See magnetic quantum number.*

Reaction field theory (electric). A theory proposed [68] by A. D. Buckingham. It concerns molecules which contain polar groups and accounts for the effect of a change in the dielectric constant of the solvent on chemical shifts. The polar groups set up "reaction fields" in the solvent which may cause a redistribution of electrons in bonds to hydrogen atoms. The effect is greatest when the dipole moment of the group involved is in a direction parallel to the bond to a hydrogen.

Relaxation. The transfer of a nucleus from a higher-energy state to a lower-energy state by loss of energy.

Rotating frame. A set of reference coordinates which are defined as revolving at the Larmor frequency.* The use of this rotating frame permits the mathematical operations necessary to examine double resonance theory.

Saturation. The equalization of the number of nuclei in the different energy states. Can be brought about by using a strong radio-frequency signal. This decreases the absorption peak height. At sufficiently high levels of radio-frequency energy, no absorption signal is observed. Nuclei in different environments may differ in the ease with which they are saturated. The signals are both reduced in intensity and broadened by saturation. Saturation must be guarded against, especially in integration.

Saturation factor, Z_0.   The factor by which the absorption signal is reduced under conditions which tend to cause equalization of the populations of nuclei in the different spin states.* For protons and other nuclei which have a spin quantum number*(I)$I = \frac{1}{2}$,

$$Z_0 = \frac{1}{1 + \gamma^2 H_1^2 T_1 T_2}$$

where γ is the magnetogyric ratio,* H_1 is the amplitude of the radio-frequency signal, T_1 is the spin–lattice relaxation* time, and T_2 is the spin–spin relaxation* time.

Screening constant (nuclear).   The magnetic field at a nucleus (H_{local}) is equal to the applied external magnetic field (H_0) minus a constant (σ) times the external magnetic field:

$$H_{local} = H_0 - \sigma H_0$$

The constant σ is called the nuclear screening constant. The screening of the nucleus is a result of the circulation of electrons in such a way as to oppose the applied magnetic field. Chemical shifts in ppm are actually differences in nuclear magnetic screening constants.

Side bands.   Two types of side bands, nuclear satellite signals and spinning side bands, are discussed in the text. Also see audio side band.*

Slow passage.   Refers to the rate at which the external magnetic field is changed with time.   The usual scan is faster than one which would be classed as slow passage.

Spin echoes.   See pulse method.*

Spin–lattice relaxation (T_1).   See longitudinal relaxation.*

Spin quantum number (*I*). A fundamental property of the nucleus
which can have integral or half-integral values. For protons,
$I = \frac{1}{2}$. The spin quantum number indicates the number of distinct
orientations $(2I + 1)$ which the nuclear magnet can assume in
an applied magnetic field. Protons can thus have $[2(\frac{1}{2}) + 1]$ or
2 distinct states. These correspond to alignment with and
against the applied field.

Nucleus	*I*	Nucleus	*I*
H^1	$\frac{1}{2}$	O^{17}	$\frac{5}{2}$
H^2	1	F^{19}	$\frac{1}{2}$
B^{11}	$\frac{3}{2}$	Si^{29}	$\frac{1}{2}$
C^{13}	$\frac{1}{2}$	P^{31}	$\frac{1}{2}$
N^{14}	1		

Nuclei with *I* greater than $\frac{1}{2}$ can also have an electric quadrupole moment.*
     The number of peaks expected in a first-order spectrum for
a group of equivalent protons coupled to a group of other nuclei
is equal to $(2NI + 1)$, where *N* is the number of other nuclei and
I is the spin quantum number of the other nuclei. This accounts
for the multiplicity rules in Table 3-3. Nuclei, such as C^{12} and
O^{16}, which have $I = 0$ are nonmagnetic.

Spin states. The distinct orientations which a magnetic nucleus
may assume in an applied magnetic field. The number of spin
states is related to the spin quantum number* and is equal to
$(2I + 1)$. For protons, $I = \frac{1}{2}$, so there are two possible orientations.

Spin temperature. See nuclear spin temperature.

Susceptibility, volume magnetic (χ_V). This is the ratio of the magnetic moment per unit volume (M) produced in a substance to the
magnetizing force or intensity of magnetic field (H) to which the
substance is subjected.

Magnetic permeability is the ratio of the magnetic moment per
unit area taken perpendicular to the direction of the flux in a
substance to the magnetizing force of magnetic field to which
the substance is subjected. The permeability of a vacuum is
unity.

Most materials have negative volume magnetic susceptibilities; these materials are said to be "diamagnetic." Materials which contain free radicals (substances with unpaired electrons) have positive volume magnetic susceptibilities and are said to be "paramagnetic." Magnetic susceptibility per gram (χ) is given by

$$\chi = \chi_V / d$$

where d is the density of the substance. Molar magnetic susceptibility (χ_M) is given by

$$\chi_M = \chi_V (MW/d)$$

where MW is the molecular weight and d is the density. See Pascal constants.*

If the diamagnetic susceptibility of a substance varies with the axis along which it is measured, the substance is said to have a diamagnetic anisotropy or an anisotropy of diamagnetic susceptibility.

$\underline{T_1}$ See longitudinal relaxation.*

$\underline{T_2}$ See transverse relaxation.*

<u>Transverse relaxation (T_2).</u> The loss of nuclear spin energy by a mutual exchange with another nucleus of the same isotope. This type of relaxation does not change the relative distribution of nuclei in the different energy states, but does lead to the broadening of an NMR peak.

<u>u mode.</u> See dispersion signal.*

<u>v mode.</u> The signal normally employed in proton work is an absorption, or v mode, signal. This signal is out of phase with the dispersion, or u mode, signal.

<u>"Wiggle-beat" method.</u> If two closely spaced signals are scanned rapidly, the ringing due to the two signals is superimposed to form a "beat" pattern. The frequency of the beating is equal to the separation of the two signals. The method is thus called

the "wiggle-beat" method.   The patterns are usually observed
on an oscilloscope.

X, Y, and Z gradients.   The axis through the centers of the pole
pieces and the sample is designated as the Z axis, the axis along
the length of the sample tube is called the Y axis, and the axis
through the sample parallel to the pole pieces is called the X
axis.   There must be a minimum of variation, or gradient, in
the magnetic field in any direction.

Subject Index

A

B

C

G

Grease, silicone stopcock, absorption in spectrum, 16, 17

H

Heteronuclear double resonance, 103
Hidden proton, location by double resonance, 102
Higher-order effects, see Spin patterns
High-resolution spectra, 2
Homonuclear double resonance, 103
Hydrogen bonding, effect on position of absorption signal, 22, 59,
     Table 2-6 (25)
Hydroxyl proton (see also Steroid)
    coupling with proton on attached carbon atom, 55, 56
    effect of exchange on signal due to, 22, 23, 55
    effect of temperature on signal due to, 59
    effect on signals due to angular methyl protons in steroids, 57

I

Impurities
    aids in identification of signals due to protons in, Table 5-2 (108)
    allowance for, in integration curve, 26, 27, 30
    complications introduced by, 105-109
    detection by use of NMR, 4, 105–109
Information obtainable from high-resolution spectra, 4–5
Insolubility of compounds, 7, 9
Integration curve
    accuracy of, 28
    aided by change of solvent, 28, 109
    detection of impurities by use of, 106
    example of the use of, 28, 29
    idealized, Fig. 2-11 (27)
    points to remember, 28, 30
    use of, 26–30
Intensity of absorption signal, 3
Intensity ratios, relative
    distortion of binomial, in first-order multiplets, 51, 77, 84
    in first-order spin patterns, 45–46
    in higher-order patterns, 77–78